RESEARCH & THE ANALYSIS OF RESEARCH HYPOTHESES

Volume 1

Kathleen Thomas Allan, Ph.D.

TO:
Antonietta Russo who has encouraged me to create this book. Also with thanks to Dorothy Shields and Amita who got me started.

539756
Library of Congress Control Number: 2014902074

ISBN:	Softcover	978-1-4931-6831-6
	Hardcover	978-1-4931-6830-9
	EBook	978-1-4931-6832-3

Print information available on the last page.

Rev. date: 10/23/2015

To order additional copies of this book, contact:
Xlibris
1-888-795-4274
www.Xlibris.com
Orders@Xlibris.com

PREFACE

The design of this book owes much to M. David Merrill, Ph.D., who in the 70's was a professor at Brigham Young University. At the time, he was researching concepts for designing curriculum. He developed a method of designing instructional materials that used the principles of Rule – Example - Practice. The term "rule" covered such items as a mathematical rule, a classification paradigm, a descriptive category, or other information that gives the students a "rule" for understanding the main concept being taught. The term "example" covered the criteria showing how the rule worked. The term "practice" gave the student a way to perform an exercise, that is, in practicing with other variations of the example to gain experience in using the rule. For Merrill's principles to be effective, the example must match the practice and the governing rule.

RESEARCH & THE ANALYSIS OF RESEARCH HYPOTHESES has been designed employing Merrill's theories. The "rule" is described and defined in the portion of each unit of instruction labeled "***Purpose:"*** ***T***he "example" and matching "practice" are described in the portion labeled "***Objectives***:" The body of the unit describes the rule in more detail giving examples as needed. Finally, there is an "assignment" which requires the student to put the rule into practice. The two volumes provide a basis for doing a research study which graduate students can use as a model for their thesis or dissertation. Volume 1 covers basic principles and processes for doing a research study and Volume 2 covers the five major procedures for testing research data: the z-test, the t-test, the Pearson correlation test, the Spearman correlation test and the Chi Square test. Note:

'Chi' is pronounced "Kai"; the 'Chi-Square' symbol is: χ^2

OUTLINE - VOLUME 1:

OUTLINE - VOLUME 2:

COURSE PURPOSE:

The two volumes of this text are required for completing this course. You will be required to practice the steps from developing an idea for your research to completing the final report. As you work through this text, your study should be a small one with a limited number of subjects and a single problem in the area of your graduate major; if you are a senior required to learn to do research, the problem can be in the area of your undergraduate major or the area you plan to major in when you enter graduate school.

Both volumes combined are designed to give each participant practice in all aspects of completing a small research study including a final report. Your report must include a section defining your objective for the research, a description of your research problem, that is, your main question, your theory about the solution, as well as why this problem is important. Your report must describe your plan for gathering information (data), how you will test your hypotheses, the results you obtain, and your report must include your conclusions and any questions the study has suggested for further research.

COURSE OBJECTIVES:

- Apply problem identification and problem solving skills in the real world.
- Define and/or describe all aspects of doing basic research.
- Identify a particular research question, form a hypothesis, and perform a literature search in the field of the research question.
- Evaluate professional literature in terms of its thesis, supporting evidence and suggestions for further research.
- Analyze the research literature that applies to your research question and write a paper describing how that research is relevant to your question and your hypothesis.
- Prepare a proposal for your research study to be submitted to your faculty advisor and department Chairman.
- Describe a research project/study to determine the validity of your hypothesis including data gathering and analysis.
- Gather appropriate data, analyze it using appropriate statistical tests described in this course, and illustrate the gathered data and analysis with appropriate graphics.
- Prepare and submit a publishable article describing your research and the results using the skills developed in fulfilling these course objectives.

TERMINAL OBJECTIVE FOR THE COURSE:

Write a report in the form of an article suitable for publication, reporting your research and the results, with a minimum standard of a B grade. Your Instructor may determine a different grade requirement.

SPECIFICATION FOR REPORTS & ASSIGNMENTS:

All Reports should be typed and double spaced. The reports for this course should be hand-in-hand with the student workbook and group discussion activities. ***Make sure spelling and grammar are correct.*** You will be required to get a minimum of a ***B Grade***. Why a B Grade? You will have plenty of opportunities to improve all reports. And to be suitable for publication any standard below B would not be adequate.

ONLINE ACTIVITIES:

Through the various sections, you will be given a selection of URLs in order to learn how to use the Internet to get information about Research, Proposals, Thesis & Dissertation Writing and Statistical analyses. Take advantage of these as the various sites (and others) will be helpful when you begin to do other studies. {Note: some of these are stable, but others may change as the owners of the site make internal changes. This is inevitable. If this happens in this course, your instructor should be able to assist you to find others. Share information with other members of the course if you find any URLs without assistance.]

RESEARCH NOTEBOOK OR DIARY:

It is really important that whenever you are doing some activity that relates to your research, that you ***keep notes of what happens.*** This needs to be a small pocket notebook not your computer! You may get insights into the research, or find some unexpected problem, or discover something you hadn't thought of. Whatever it is, if you don't keep notes you will not remember the information that may add a special insight to your research paper. ***So...Get in the habit of carrying the research notebook/diary with you all the time. And USE it!***

ASSIGNMENT WORKBOOK/COMPUTER FOLDER:

You need to keep your class notes, assignments, etc., in a workbook for this class or in a folder on your computer (keep it up to date). Always start a new page for each assignment. You will often be required to prepare handouts for the rest of the students in your class so they can review your work as part of a class discussion. Arrange for copying with your instructor.

- One major purpose of the group discussion is to enable each person in the class to improve their work.
  - When critiquing, be honest but not unkind.
  - No put-downs or derogatory remarks.
  - Be helpful with suggestions for improvement.

Table of Contents

SECTION 1: PRINCIPLES OF RESEARCH.

PURPOSE:

Research isn't just a matter of looking something up, writing a description of what you found, and giving your opinion of the effects and/or benefits of your findings. While those may be elements of research, what is needed from seniors and graduate students is something a lot more formal; something that is guided by a specific question or set of questions in a particular educational field of study. Formal research is the process of collecting, organizing, analyzing, and interpreting data for the purpose of gaining insight, increasing understanding of the research question, and presenting the results to our colleagues. Your ultimate aim is to satisfy not just yourself but others in the same field as yourself. Your questions must address concepts or ideas that will benefit or improve practices in your field of study and assist in increasing understanding of the principles and standards within your field.

The first vital component in doing research is to have a good, working plan. There are a lot of elements required for a quality plan. You must properly define the problem you are interested in and define any specialized terms that are essential to the area of your study. In addition, you must be able justify your investigation of this particular problem. You must determine the methodology you will use to complete your investigation (what you will do and how you will do it). Part of determining the methodology requires knowledge of populations and samples. You must also be certain that the data you collect will enable you to reach valid conclusions. In order to do that, you need to understand the nature of data, that is, what types of data may be sought, how they are similar and how they differ. Then you must convert your problem definition into a hypothesis, an explanation or theory of why the problem exists. The next stage is gather the data and analyze it to determine if your hypothesis does explain your question.

The major purpose of doing research is to add to the body of knowledge for your major field. What this means is that the data you collect must be analyzed and, yes, it does use a tool called statistics. But it is simply a tool. Once analyzed, conclusions are drawn and you write your final report. All these activities are an integral part of the research process.

OBJECTIVES:

- Determine what **research** is and why it needs to be done.
- Define the meaning of **population** and **sample** as far as research is concerned.
- Describe the different types of data that can be gathered in a research study.
- Describe why it is necessary to analyze the data that is gathered in research studies.
- Describe the basic methods of doing research, particularly in the area of the educational and social sciences.

A BRIEF OUTLINE OF THE RESEARCH PROCESS:

Developing a research question leads to formulating the problem statement. Then you will need to decide what research method to use. Who will be the subjects in your research?

This will require an understanding of the population from which you will select your subjects and the best size of the sample you will select.

You'll also need to understand the nature of data and what type of data will be of most value in solving your problem. You also need to find out what has already been done in the area of your problem statement which means doing a search of the literature in journals that cover your field of study.

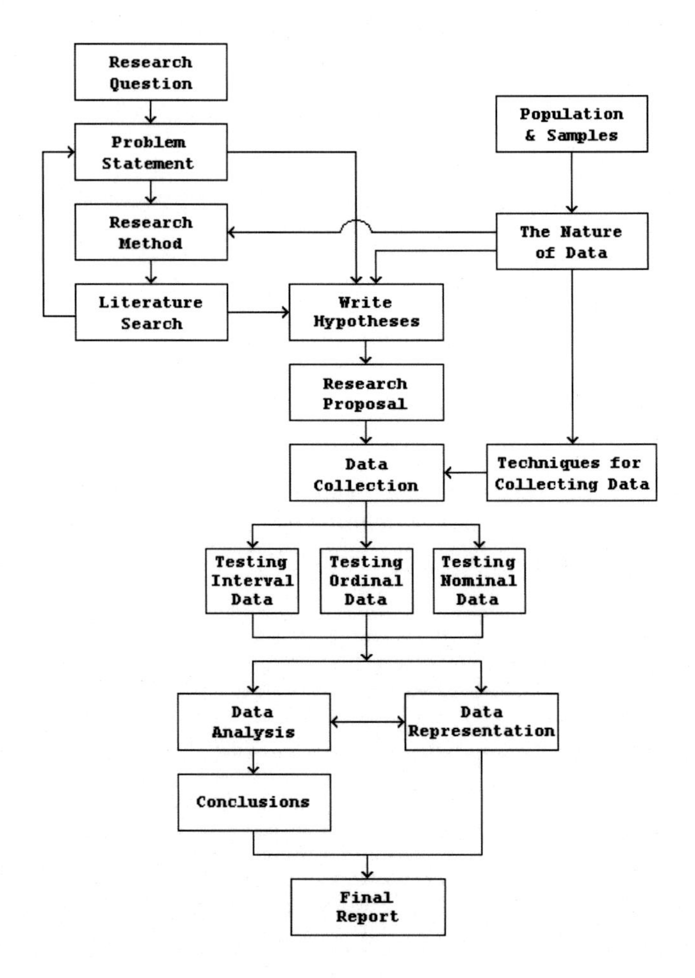

Figure 1: The major concepts required for performing a research study.

You must convert your problem statement into one or more hypotheses to present your theory about the questions you have. They guide your study and the tests you will need to complete in order to determine if you can accept your hypotheses or must reject them as untrue. You will have to collect appropriate data; the test of the data is what helps in the acceptance or rejection of your hypotheses. You must examine the results of your study and form conclusions which should be a reasoned analysis of your study.

It is helpful to create some graphics which will represent the data and the results of your study. This is combined into your final report. The report presents the results of your study to your colleagues and hopefully will expand the knowledge base in your field.

UNIT 1A: WHAT IS RESEARCH & WHY DO IT?

PURPOSE:

UNIT 1 will help you start work on your research plan, which will be based on a combination of: the purpose for your research (what you aim to achieve); the information in this unit; and the list of steps you need to follow. Note that the step you will complete in Unit 1, Sections A-E, is step 1: Determine what question in your field you would like to study and write the question as a problem statement.

The end product of Unit 1 will be your problem statement. A group discussion will provide an opportunity to improve your problem statement and assist others in the class with a critique so they can improve theirs.

OBJECTIVES:

- Review the basic characteristics of research described by Paul D. Leedy, 1974[1] and select an area of research from your planned graduate major that you are interested in studying as part of this course.
- Prepare a *Research Question* (maximum of one double-spaced page) describing your research interest, what you already know about that interest and why you feel it would be useful to yourself and others to find out more.
- Make your research question into *a clear statement of the problem.*
- The statement should be as simple and as concise as possible but do not simplify at the expense of clarity.
- As this problem statement will determine the direction of your research efforts this semester, you must have a clear conception of what you are going to do. And starting with a clear statement is the first step in writing your final report.
- Include a description of the variables or contributing problems that help define the research question you have identified.
- Participate in a group discussion to critique each student's Research Question from the viewpoint of scope and focus (is the question too broad or inadequately defined).
- Use the groups' critiques to rework and improve your statement.
- Hand in your original problem statement, the critiques from the other student and your final problem statement.

---

1 Paul D. Leedy, Practical Research: Planning and Design. Macmillan Publishing Co., Inc., 1974. ALSO: Paul D. Leedy & Jeanne Ellis Ormrod. Practical Research: Planning and Design, Ninth Edition. © 2010, 2005, 2001, 1997, 1993, Pearson Education Inc., publishing as Merrill, Upper Saddle River, New Jersey 07458.

INTRODUCTION:

This section provides information on what research is about. Your first step is to properly define the problem you wish to learn more about and define any specialized terms that are essential to your field of study. These terms may be so much a part of your understanding of your graduate major, that you may not realize they need to be defined for your potential readership who might have no more than a basic understanding of your field. There will be many among your associates who already know the terminology, but it is better to define the terms, because you may be using them in a slightly different context and your definition may provide a new "light" on the subject; at the very least, your readers will know how you are using the terms.

For example, the author was looking at two methods of calculating the standard deviation and realized the only difference was in the way the variance was calculated. Oh, but wait a minute – you may not know what the two terms standard deviation and variance mean! These terms will be defined later in this course. But this is an example of how the author's familiarity with the terms could leave readers wondering what is being talked about. The same applies to your field of study.

LEARNING TO DO RESEARCH:

First, for much of this course you will be learning to do research with a do-it-yourself hands-on approach. There will be a number of occasions when you will work with other students in the course who may or may not be in a field different from your own. The basic principles of research apply to just about every field in the social sciences and in the physical sciences. For example, an associate does research into taste tests of edible products she is involved in developing for her company. This may appear to be unusual, but no matter what your discipline, this course will enable you to learn how to conduct research.

Second, the experience of working with a discussion group will enable all in the class to get different viewpoints, to so-to-speak "wrestle" with the activities you will be working together on. The outcome of these discussions and evaluations of each other's work will enlarge your understanding and will enhance your written products.

THE RELATIONSHIP BETWEEN RESEARCH & STATISTICS:

Let's repeat the definition of research given earlier: Formal research is the process of collecting, organizing, analyzing, and interpreting data for the purpose of gaining insight, increasing understanding of the research question, and presenting the results to our colleagues. The definition of statistics is similar: Statistics is the science of collecting, organizing, analyzing, interpreting and presenting data for the purpose of gaining insight and making better decisions. There are five different types of analytical tests. The analyses you will need to do will NOT be complex. Each test has been formalized as an algorithm or set of rules. Each algorithm is designed to be as simple as possible and is a step-by-step procedure for completing the analysis. All you will need to do is follow the steps and you should have no problems with the calculation.

In this course, the study should be a comparatively small one. When you do your Thesis or Dissertation, you will be doing a larger research study. The Thesis is generally based on research that has already been done; the Dissertation is, while related to previous research, generally also breaks new ground. The principles and procedures taught in this course will transfer to those studies. You may need to get the help of the statistics department if you need to use the ANOVA program (which compares three or more variables).

You will not be alone. Helping you to do the analysis will be at least one statistics graduate assistant. Your study's analysis will be good practice for him or her, and it will take the analytical burden off your shoulders. Treat the graduate student(s) as your expert (they will be) but you still must know enough to be able to direct them. For example, when one graduate was ready to have the results of her dissertation analyzed, the graduate student wanted to analyze 10 variables. She had to redirect him and explain there were only four variables of interest in the study.

CHARACTERISTICS OF RESEARCH:

Although the book this part of the text is based on was first written so many years ago, the description of research still deals with essential aspects of the research process. Leedy presents several major elements or characteristics of research listed below.[2] Later on units in this textbook will provide more details on each concept:

1. Research begins with a question in the mind of the researcher.

Leedy indicates that the world around us is filled with unresolved and often baffling situations. Because people are critters filled with curiosity, we endeavor to determine the meaning, the cause and/or the effect of such situations. Our curiosity leads us to formulating questions that can become the basis for research.

Your research question must be a clear statement of the problem you are interested in studying. In order to provide the goal for your research, your problem statement must be clear and simple, written in precise terms and define what you are seeking to discover. A clearly written problem statement will allow you to stand back and look at your goal objectively, an essential step in planning.

2. Research may require the development of a theory.

Developing a theory is particularly important where little or no research has been done. As you get more experience, you may come up with new ideas about some aspect of your research field. Developing a theory may be a necessary part of planning larger research studies. The small study you will complete may be regarded as your personal theory development. The researches you complete in the future will be enhanced when you ensure you understand the various facets of the research process.

3. Research requires a plan.

The plan can only begin once you have settled on what your research question will be. The goal must be firmly fixed in your mind at all stages of your study as this is what gives direction to your research. For valid answers to your question, your research must be planned and designed carefully. Research is a process so all necessary elements must be present and the design of the study must follow a logical pattern. Research demands a clear statement of the problem & deals with the main problem through supporting concepts.

Many researchers fail to isolate the sub-problems that are intrinsic to the problem area on which they are focusing. As a result, their research plan becomes unwieldy and/or poorly designed. They will miss

2 Paul D. Leedy & Jeanne Ellis Ormrod. Practical Research: Planning and Design, 9th edition. Pearson Education, Inc., publishing as Merrill, 2010.

important information that could mean having to do the study again to include what was missed. In such circumstances one cannot be certain of the validity of the results and conclusions. For example, one researcher focused on characteristics of capital letters in the way a child learned to recognize them. These characteristics were: straight vertical lines, straight horizontal lines, and straight slanted lines in either direction from the vertical. One of the conclusions of the researcher was that he ought to have included open and closed curved lines such as those for B, S, C and O. A bit of forethought and an examination of the set of capital letters would have avoided that error.

Your aim will be to examine your problem statement and define all the lesser problems or supporting concepts. When you have resolved the lesser problems, together they will provide the solution to your main problem. However, for this course, your study will be a small one, with probably no more than 10 subjects and no more than one or two supporting concepts. If you find there are more sub-problems, you may need to refocus your goal so that you have a single question to be solved. Leave the other questions for later research.

4. The direction of research is defined by appropriate hypotheses:

For each supporting concept a hypothesis needs to be formulated. The hypothesis is an educated guess or a logical assumption or a theory concerning the possible solution to the stated problem. The hypotheses give researchers direction for gathering data, the type of data that needs to be gathered, and what analyses are most appropriate for that type of data.

To provide a focus for your research, you will need to convert your problem statement or question into a hypothesis. You need to have an idea of what the answer to your question could be, a "theory" of what might result from gathering data, an educated guess as to what the solution to your problem is likely to be.

Research deals with facts and their meaning and describes any assumptions made: Having stated the problem, isolated the supporting concepts, and formulated your hypothesis, the next step is to collect and examine the facts, concepts, and measurements of data that seem pertinent to the problem, organize them in a systematic way, and interpret the results. There are a number of concepts that you will need to understand, such as populations, samples, the nature of data, and how the data may be analyzed.

You may have to make some assumptions, that is, there are concepts that seem to be self-evidently true. It is essential to identify the assumptions you make so that, in your final report, you can describe them and explain why you feel it is appropriate to assume that "this is so."

5. Research is circular:

Often, our research, while answering some questions, may provide us with one or more new questions, or may open the door to another area in which we can perform some research, in which case we begin the whole process again. A single study is also circular: Define the problem and formulate a hypothesis; the hypothesis directs the gathering of data; the analysis of the data provides a means of validating the hypothesis and may lead to redefining the problem or provide a new or extended problem.

RESEARCH OBJECTIVES:

The purpose in doing any research is very important. You need to know what you plan to do and why you are doing it. In fact, most research questions can be put into one of three categories.

1. Determine "how things got to be the way they are."

Answering a research question relating to "how things got to be the way they are" is generally descriptive. The researcher is dependent on records provided by others, since many of the events that have led to the current situation are in the past. Findings will be more valid the closer you get to the source, such as, eyewitness accounts, autobiographies, accounts of acquaintances of the individual concerned, an interview with the individual. If the researcher can pinpoint the events and/or research that led to the current understanding of the situation, it is possible to question any "automatic" conclusions.

2. Describe characteristics and determine "what seems to be so."

Other research questions try to determine what seems to be so. This type of research is both observational and descriptive. The researcher observes the situation and describes the characteristics found.

3. Test a theory about "why it seems to be so."

The third category of research questions are concerned with why things are the way they are. This category deals with cause and effect, and a related question may be *what can we do to change, enhance, or remove the effects?* In research of this type, theories are formed to account for observed phenomena, and the theory is tested in some type of an experiment. The researcher seeks to control what is occurring in order to determine what is affecting the observed phenomena.

BEGIN THE PROCESS FOR YOUR RESEARCH STUDY:

Use the following points to help determine the question you would like to study and write it as a problem statement. Use all the information given in this unit to assist your thinking.

- What are your major interests in your field?
- What do you want to know and/or find out?
- Why do you want to research this question?
- Justify your research by explaining why it is of value to yourself and others in your field.
- What terminology is likely to be unfamiliar to people outside your field of study, including the students in this course? NOTE: Better to start with too many terms rather than too few! You can always edit out unnecessary terms.

UNIT 1A Assignment:

1. **Write a short summary of the characteristics of research.**
   - **Write up your problem statement (no more than a double-spaced page);**
   - Include why you wish to study the area the statement defines;
   - Identify and define any terminology specific to your study area.
   - Bring enough copies of your problem statement with your name on it for each member of the class.
2. **Provide the rest of the group with a copy before the discussion.**
   - Write up critiques with explanations (what could be improved and why) for each of the other students' problem statements.
   - NOTE: Each critique should be no more than a double-spaced half-page.
   - The critiques should be designed to help the other students (no derogatory remarks or put-downs; make the discussion a pleasant experience for all).
3. **Discuss what you have learned with the other group members.**
   - Come to a consensus concerning how the concepts in this unit may be useful in defining and working on your research studies.
   - During the discussion provide feedback concerning the critiques on your problem statement and the problem statements of the other students.
4. **Revise your problem statement and hand in:**
   - The critiques of your paper.
   - The original and, clearly labeled, your revised problem statement.

UNIT 1B: POPULATIONS & SAMPLE SIZES.

PURPOSE:

Before you get too far into your research plan, you will need to convert your problem statement into a hypothesis. One purpose of the hypothesis is to direct your research concerning the number of subjects you will obtain data from and a definite description of the type of subjects you will need. For example, will you prefer to work with adults?, Children?, Teens?, Or will it be with animals?, Bugs?, Or other flora or fauna?

Note that there is a two-way interaction between these factors and your hypothesis. Before you can construct a quality hypothesis, you need to know what a population is, what a sample is and how you get one. It is from a population you draw your sample so it is necessary to define your population very specifically (how big a group will you need?) otherwise the sample you draw may not be a good fit to the population you are considering. For example, if you want to work with third graders or kindergartners, it is no use focusing on a population from the local middle school or junior high. The composition and size of your chosen population determine the composition and the most appropriate size of your sample.

OBJECTIVES:

- Define the terms "population" and "sample size."
- Describe the three factors that ensure that the size of a sample is an effective one for a given population.
- Describe the population from which *your* study will get samples and determine the best size of sample for your research needs (how big a group will you need?).

INTRODUCTION:

When doing research, we need to look at an appropriate population for the gathering of data. If you were doing research into the composition of the families in a small town, for example, it wouldn't make any sense to look at a population comprised of single people. Because any population may be of humans, animals, samples for a production line, or anything else, each member of the population or sample may be referred to as a unit of observation or a unit of measurement. When you gather your data, you are measuring and/or observing those in your population or sample.

WHAT IS A POPULATION?

A population is a collection of units of observation (members or subjects) that a person may be interested in studying. The population must be well defined so that it is a simple matter to distinguish between members and nonmembers. Examples of populations are:

1. All members of the human race.
2. All students at Valley View High.
3. All first year students in Southern Illinois Medical School.
4. All kindergartners in Vineyard Elementary.
5. All the Lepidoptera (moths and butterflies) in the grounds of Thanksgiving Point.
6. All the plants growing in one square foot of ground in your backyard.

CHRACTERISTICS OF POPULATIONS:

Two of the major factors that relate to populations are size and composition. Notice in the above list of populations, the size of the population decreases from item 1 to item 6. Most studies are done where the population is finite, in other words, limited in size. The population of the world is a finite number, that is, there is a limit to the number of people, but that number is constantly changing. For all practical purposes, 6,000,000,000 (six billion) or more might as well be an infinite number.

Another example of an infinite population would be all the sand grains on all the beaches of the world or the grains of sand in the Sahara Desert. While there is a finite limit to the number of grains, the task of counting them would be worse than daunting! How would you select a sample? About all it could be would be a tin bucket full of sand? And what would the problem statement be questioning? Other "infinite" populations with similar problems could be "all the asteroids in the solar system" or "all the citizens in the United States".

Any size population from ten to hundreds may be used in a research study as long as you are very specific in its description. For example, saying "I am going to study high school students" is not specific enough. Even adding "at Valley View High" still doesn't add enough specificity. Are you going to study the whole population of students at Valley View High? What are the characteristics of the people who live around Valley View High? Is it a rural community, in the middle of the city, or in the suburbs? Is it a young community or a dormitory town or mostly retired? What size is Valley View High: less than 100 students,

500-600, more than 1,000? And so on...

The description of the population determines what characteristic or characteristics you could study. Determining why you want to study that population can only be done if you know that 'what you want to study' is in the description of the population you are looking at.

GENERALIZING STUDY RESULTS:

Since your study is to be a small one, the population you look at should be fairly small. For example, if you want to find out something about kindergartners, the population of kindergartners in a single school would probably be less than 50; in a single classroom, perhaps 20 or 25. Of course, if you select a single classroom in a specific school, your study will tell you about the kindergartners in that classroom or school only. You will not be able to generalize your findings to other schools, even if they are in the same district.

Why is this? It is the other factor mentioned: composition. Trying to compare an East coast city with a small town in the Dakotas, you can see that the composition of the two populations would be very different. In other words, there is diversity between the two populations. Because the two populations are so different, it is unlikely that you would be able to get meaningful data. As a general rule: the greater the diversity between two different populations or the greater the diversity within a population the less likely you are to get a sample that is a good representation of the population.

USING THE WHOLE POPULATION:

A solution to the diversity problem is to use the whole population. Then we can be 100% sure that as we measure a particular attribute or characteristic of the population, our results represent reality with regard to that population. Suppose we were interested in the students at a particular high school. If we obtain data from every single student, we know "all there is to know" in the area we are investigating, provided we asked the right questions.

Using the whole population being investigated in our research, we can be reasonably sure that our information is accurate *with reference to that particular time **and** under the same circumstances.* A different year changes the composition in that students in the senior year have moved on and new freshman students have entered the school. Even if one were to have the exact same population, an additional year may change the mix and/or attitudes in the population.

Even if we use the whole population at one high school, trying to generalize the information we obtain to "all students in all high schools" won't work, not even if we consider only the high schools in a single school district. It would not be possible to generalize unless we could prove that all high schools in the district were identical in composition.

An example of trying to generalize beyond the population under consideration: A few years ago, a report on the "Mormons" of Utah showed a much lower incidence of certain cancers than the general population in other areas of the U.S., the inference in the report being that the way of life of the Mormons affects the incidence of cancer. But if one were to compare the whole population of Mormons in Utah to the whole of the U.S. population, such an inference is not necessarily accurate owing to the diversity of the subjects in the two populations. In addition, the difference may be attributable to the mountainous location rather than the life style. To be more sure that the inference is correct, the study should compare Mormons with non-Mormons living in Utah. Or Mormons in Utah compared with Mormons living in the rest of the U.S. In addition, the subjects of the study in Utah should be compared with other mountain states populations, where there are also fairly large numbers of Mormons. Again, Mormons in each of the mountain states could be compared, as well as comparing the Mormons and non-Mormons who lived in the other mountain states.

As a general rule, the closer in size, living conditions, and other characteristics between and within samples, the more easily comparisons can be made that give valid results.

CERTAIN FACTORS MAY AFFECT SUBJECTS IN YOUR STUDY:

Even when we work with the whole of a population, we are unable to control all the factors that may affect the results. Some of the factors that contaminate the results of a study include:

- MATURATION:

  People's knowledge, attitudes, and level of experience change even over short periods of time.

- MORTALITY:

  Some of the members of the population "die out" that is, are removed from or remove themselves from the population being considered.

- INSTRUMENTATION:

  Your method of collecting data may change over the duration of the study. For example, as you obtain responses when interviewing, you may consciously *or* unconsciously change the questions being asked in reaction to those responses.

- EXTERNAL ENVIRONMENT:

  Concurrent influences not under your control may affect the population during the course of the study. A simple example: A T.V. Documentary may be aired that relates to the area you are studying.

What is important is that all these factors may affect the results of the study. This means that when doing research, we must be aware that our study will not provide all the answers and there is also the possibility that the answers we do get may not be right. If this is true of a whole population, what if we are working with only a small subset of the population?

POPULATION VERSUS SAMPLE:

When the size of the population is large enough, practical ways of studying the whole population are not available. A researcher cannot wait, for example, for all light bulbs to burn out before coming up with an estimate of how long, on average, a particular type of light bulb will burn. It is impractical to obtain data from all locations in the U.S. In cases where it would be unreasonable to obtain data from a whole population, we select a sample. The most objective way to obtain a sample of the population is to use a table of random numbers or a computer routine that is programmed for random selection. In doing this we hope that our sample will be representative of the population from which the sample is to be drawn.
Generally when doing a research study we are looking at some attribute or attributes common to both population and sample. To distinguish between the population and the sample, we use the term ***parameter***

when referring to an attribute of a population and ***statistic*** when referring to an attribute of the sample. This distinction is made even when the characteristic we are looking at has the same description for both population and sample.

The same factors described earlier (maturation, mortality, instrumentation and external environment) can affect the sample as well as the population; but there is an additional factor that applies to the sample, that of ***chance***. When we take a random sample, there is a small but definite probability that the sample we obtain will not be truly representative of the population with reference to the attribute we are interested in.

For example, suppose you draw a sample of 20 from a particular population and find all members in the sample are red-heads. This sample would only be representative of the population if the whole population you were considering also had red hair. It would be a matter of chance, if your selected sample consists of only red-headed people, when all colors of hair are found in the specific population.

To be useful, your sample must be a realistic cross-section of the population. The larger the sample we draw the greater the probability the sample will represent the population. To emphasize the difference in collecting data for a population versus a sample, in a population we collect all possible responses, measurements or descriptions relating to the question of interest. In a sample, we collect the same type of data but we are working with smaller numbers of observational units. The sample is a subset of or belongs to the population we are studying, so how large should our sample be?

SAMPLE SIZE:

When testing a sample, the assumption is made that the sample is truly representative of the population from which it is drawn. If the population is small enough, we can use the whole group but practicality generally requires us take a subset of the population. The size of the sample depends on the size of the population. An appropriate size for our sample depends on several factors: the size of the population, how confident we want to be that the sample resembles that population, how much variability there is among members of the population, and the method by which the sample is selected. Fortunately, the work of determining sample size has been done for us. Below is a sample-size table for specific finite populations. Even if you are faced with a very large population, there is a limit to the size of the sample you need to select.

POPULATION SIZE	SAMPLE SIZE	POPULATION SIZE	SAMPLE SIZE
10	9	1,000	277
25	23	2,000	322
50	44	5,000	356
75	62	7,000	364
100	79	10,000	369
200	131	50,000	381
300	168	70,000	382
500	217	160,000	383
750	254	1,000,000	383

Figure 2: Standard sample sizes for different populations.

FACTORS USED IN DETERMINING SAMPLE SIZE:

If the population size is 10, we must use at least 9 members of that population, and if 25, the sample size is 23. When working with a population of not more than 30 units, it is more practical to use the whole population. At the other extreme, a sample size of 383 is all that is needed if we have a population size of 160,000 or more. For the information in Figure 2, three specific factors were used to determine how big the sample size should be for each given population size. These are: degree of accuracy, proportion of sample size, and level of confidence.

- <u>Degree of accuracy = ± 0.05</u>

  When your data provides an average for the sample in your study, this average is referred to as the ***sample mean.*** Because the sample is only an approximation of the total population, this mean can only be an estimate of the population's ***true mean***. The ***degree of accuracy*** indicates how precise this estimate is. A value of ± 0.05 indicates that the value of the sample mean is within 0.05 points above or below the true population's mean. For example: Suppose the sample mean is 15, then, with this degree of accuracy, the true population mean would lie somewhere between 15 ± 0.05, or between 14.95 and 15.05. If the population mean was 12, the sample mean should fall within the range of 11.95 and 12.05 of the true mean. If it is not within this range then the mean must be tested to show whether the sample mean is significantly different from the population mean.

- <u>Proportion of the sample size = 0.5</u>

  This item refers to the amount of prior knowledge one has of the population with reference to the attribute of interest. By definition, the measure of 0.5 indicates there is no prior knowledge. The assumption in Table 1 is that with the indicated sample size for any population value, there is no prior knowledge.

- <u>Confidence level = 95%</u>

  This item indicates how much confidence you can place in the sample being a good cross-section of the population. A 95% level of confidence indicates that if you selected 100 different samples from the same population, 95 of those samples will truly represent the population, and five will probably not.

DECIDING THE SIZE OF *YOUR* SAMPLE:

The study you will do for this course should be kept small. This is partly to keep the work you have to do to a reasonable amount in order to complete the study in one semester. And partly because for a first-time study, you will gain a better understanding of the research process if you keep it small. It is recommended that you have no more than ten (10) or fifteen (15) subjects in your study. "Subjects" are often referred to as units of measurement.

Note that the parameters or characteristics of the population must be defined exactly before drawing a sample. Unless you know exactly how your population can be described, you cannot be sure your sample is truly representative of that population. Assuming your sample is a good representation, the conclusions that you come to about your sample can be generalized to this specific population but not beyond to any other population, not even one that has similar parameters.

SELECTING YOUR SAMPLE:

The closer to a random selection you can get, the more likely your sample is to accurately represent your population. The purpose during selection is to try to reduce bias. In this context, bias means the introduction of some kind of distortion that will affect the results of your analysis. If the results of your study are distorted in any way, your colleagues will not be able to have confidence in your conclusions.

If you take all the kindergarten children in a single class, the whole class becomes your population. No need for random selection. However, if you are selecting a sample from all 175 kindergarten children in a school and only need 25 children for your sample, they should be randomly selected. In preparation for selection, you need a list of all 175 kindergartners, with each child being assigned a number between 1 and 175, no duplicates.

Putting the names in a 'hat' and pulling them out after shaking, or tossing a coin or dice are not very good methods for randomly selecting the members of your sample! The ideal method for selecting a sample from a population is to generate a set of random numbers. You need to use some form of a random number table to generate the numbers. When a number is generated, the child with that number is part of your sample. You continue to generate numbers until you have the number of children you need for your study.

Suppose your population is "all the kindergarten children at Sunset Lake Elementary" and suppose there are 100 kindergarten children at the school. The simplest method is to list all 100 children alphabetically regardless of which class they are in and assign the numbers beginning with one for the first child on the list and continuing through the list to one hundred. Then, when a number is generated, the child with that number is part of your sample. You continue to generate numbers until you have the number of subjects you need for your study.

Check out this Internet URL: http://www.random.org/
Looking at the generator (see Figure 3) at this site, there are two fields labeled Min and Max. The default numbers are 1 and 100, but these numbers can be changed to fit your needs. Below the two fields is a button titled Generate. Pressing this button gives you a single number. You can press the button as many times as you need. This particular generator is truly random as it will sometimes generate numbers that are duplicates of earlier generated numbers. For example, on one occasion, 34 came up twice.

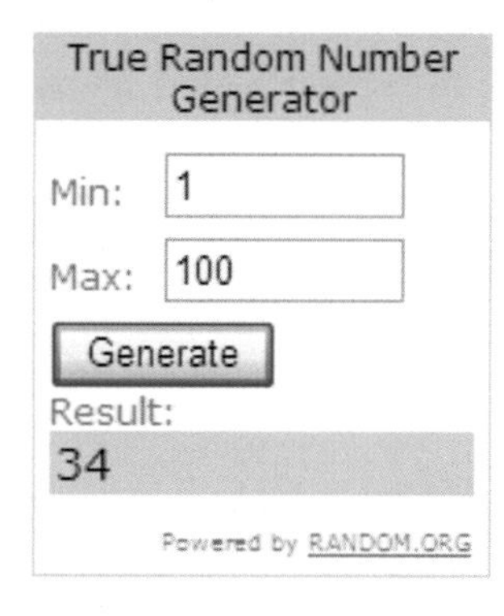

Figure 3: A generator of truly random numbers.

At least one site allows you to turn off the possibility of having the same number appear more than once. However, for true randomness, the numbers should allow for duplication. Then again, since we don't want to have the same person twice in our research study...

The random number generator shown above generates a single number at a time. If you type, in the Google search field, the following words: "random+number+generator" (no spaces) you'll get many listed. Some allow you to generate a single number at a time, others provide you with a whole set. Try out several of the internet sites.

The URL for the site for the example below is:

http://www.psychicscience.org/random.aspx

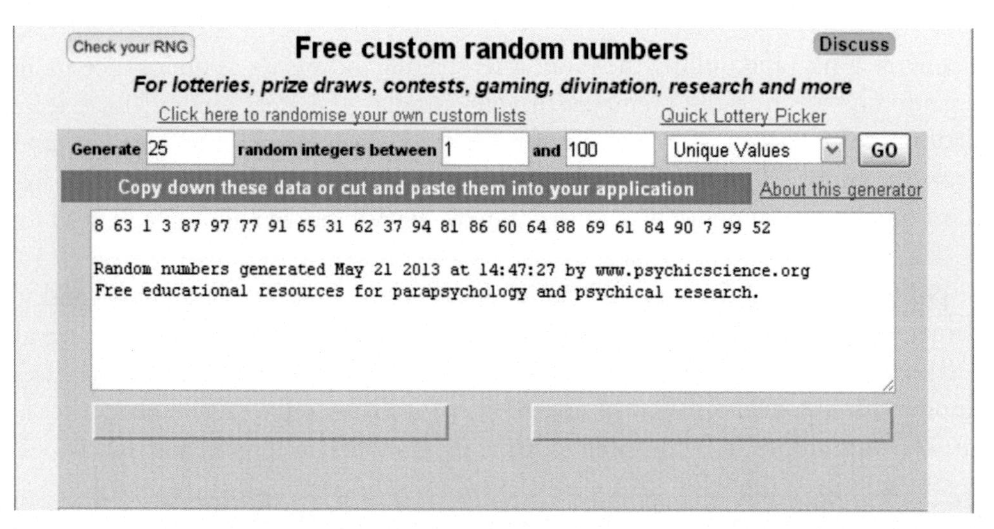

Figure 4: A random number generator that generates 25 or more numbers at the same time.

Write the quantity of numbers your need beside ***Generate***; insert the range of numbers to be selected from, in this example the numbers 1 and 100 were used; and if you need an exact number of subjects in your sample, you should change the "Open Sequence" to "Unique Values." Once you have your randomly generated numbers, pick the children whose numbers on your list correspond with the set of random numbers. For example, using the generated list, you would select the children whose numbers were 8, 63, 1, 3, 87, etc.

Selecting numbers randomly in a research study helps to decrease the chance of bias in your sample, something you always need to watch out for.

UNIT 1B Assignment:

1. **Write a report of no more than two double-spaced pages:**

a. Given your problem statement, describe the population's characteristics, including its size, from which you will draw a sample and the best size of the sample for that population (keep the sample small for this study!).

b. Describe the three factors that help determine if the data from your sample will be valid and meaningful.

c. Prepare copies for the rest of the members of the class and put your name on it.

1. **Critique the papers of each member of the discussion group (write notes on the paper for the student's information; make sure it is helpful).**

2. **Revise your report and hand in the critiques of your paper, your original and your revised report.**

UNIT 1C: THE NATURE OF DATA.

PURPOSE:

The size of your sample is not the only variable you must take into consideration when planning your research. The major tool for answering questions based on your problem statement is data. [Data is the plural form for items of information, the singular form is datum. However, in this course we will use the term "datapoint" for a single datum.] Data are the information we collect for the purposes of our research.

The nature of the data you are planning to collect will have significance in both the collection and analysis stage. In fact, the type of data you collect will determine what statistical tests you are able to apply, and the nature of the data you plan to collect will have a strong impact on the wording of your hypothesis.

Plan to collect all three types of data, interval, ordinal, and nominal, for the study for this semester. You need the experience of figuring out how to collect and analyze each type.

OBJECTIVES:

- Define the terms nominal, ordinal and interval used to describe the type of data that may be gathered.
- Given a set of data, determine whether it is nominal, ordinal, or interval data.
- Given a brief description of several research questions, determine which data type (nominal, ordinal, or interval) will be collected.

INTRODUCTION:

Information (data) gathered for a research study must be organized in preparation for analysis. Basically there are two categories of data. One is ***quantitative***, measured by numeric values. The data are numerical and provide opportunities for calculations using addition, subtraction, multiplication, and division. The other type of data is ***qualitative***, that is, such data is described by qualities or characteristics and is non-numeric. The data is non-parametric, measured by classification only because the data can be classified by its characteristics.

PARAMETRIC & NON-PARAMETRIC DATA:

Parametric tests are used on data that describes a property of the population that is quantitative; the property is referred to as a parameter. One or more parameters show the limits or boundaries of the population. The sample you select must exhibit the same parametric limits. Quantitative measurements come in the form of numbers attached to a unit of observation and show capacities (frequencies, answering the question "how many") or dimensions (time, age, distance, etc.), and are measured on a naturally occurring scale. For example, time is measured in seconds, minutes or hours; distance is measured in feet, yards, miles or centimeters, meters, and kilometers. The scale used may be based on the set of whole numbers, a number line beginning with zero, or the real number line where zero is the central or pivoting point; positive numbers appear to the right of zero and negative numbers appear to the left of zero.

Non-parametric tests are used on data that do not have quantitative information. This type of data is called nominal where only certain attributes or characteristics of the population/sample are being measured. The categories cannot be ordered but supply information such as gender, zip code, color, etc. Nominal data may appear to be numeric as in the zip code *but the numbers are assigned arbitrarily and used as a label.* If the information is categorical but can be ordered, it is called ordinal. The data gives a ranking such as the sequence of people crossing the finish line in a race, first, second, third, etc. Rankings also apply, when opinions or preferences are obtained, such as "like very much" through to "dislike very much." Preferences are solicited using numbers (1 through 5) for items such as cookies or candies, and often there are more than five items to choose from. But it is important to understand that these numbers cannot be treated as arithmetic values, etc., as they are only labels.

USEFUL DEFINITIONS:

Class:

Class is from the word classification. When data is collected and placed in a particular category, the category is a class. When data is classified in groups, such as grouping by ages: 0-4, 5-9, 10-14, etc., each grouping is also a class.

Class Interval:

Numbers defined arbitrarily by the lowest and highest numbers in the class; for example, 10-14 is defined by the lowest, 10, and the highest, 14, that is, both "ends" of the interval must be included. The range of the interval is five (10, 11, 12, 13, 14) not '14 minus 10' which totals to four.

FREQUENCY:

The number of times a particular value or datapoint occurs in a single class or in a class interval.

FREQUENCY TABLE:

The frequencies of a set of data organized in table format.

STATISTICAL INFERENCE:

A statistical inference is a conclusion about an entire population drawn from data taken from a sample.

QUALITATIVE DATA:

Measures of characteristics, traits, or qualities associated with a unit of observation, such as race, gender, hair color, political affiliation, brand name, place of birth, etc. You might refer to them as descriptors since the characteristics, etc., describe the qualities of the data.

QUANTITATIVE DATA:

Numbers obtained from measurements made on a unit of observation, that is, quantities or amounts such as height, weight, size of family, volume, area, speed and so on.

THE NATURE OF DATA:

The nature of the data you are planning to collect will have significance in both the collection and analysis stage. Data may be classified as one of four basic kinds: nominal, ordinal, interval, and rational. The usual types of data collected by those doing research in the Social sciences are nominal, ordinal, and interval. Nominal and ordinal are non-parametric data. Interval data is parametric.

In the figure on the next page, a continuum is a continuous sequence of data. The data may be discrete (individual and distinct) or it may be on a continuous scale such as the real number line. Rational data often includes an absolute zero. One use of "absolute zero" refers to the Kelvin temperature scale where the absolutely lowest temperature in the cosmos is zero (theoretically that's equivalent to -273.15°C). In other words, "absolute zero" is a fixed temperature, unchangeable and represents the lowest possible temperature in the whole universe.

DATA TYPE	NATURE OF THE DATA	DESCRIPTION
Nominal	Categories	Classification by attribute
Ordinal	Ranked Order	An ordered sequence
Interval	Numeric values	Values falling in a continuum
Rational	Rational values	May be in the form of a ratio or include an absolute zero value

Figure 5: The four types of data.

Zero is used in other temperature scales (degrees Fahrenheit or Celsius) but the zero is not an absolute value. It has a relative value depending on the scale or continuum being used. There is the Celsius scale, with 0°C representing freezing point and 100°C representing boiling point, a range of 100°C. However, the Fahrenheit scale is different from the Celsius scale. Indeed, 0°F is not equal to 0°C although one can be converted to the other. The equivalent temperatures in Fahrenheit are 32°F and 212°F, a range of 180°F. On both scales there are temperature readings that are below the freezing point. These are given negative values. Let's consider the four types of data:

NOMINAL DATA:

Nominal data consists of classifications and categories, such as, male/female or yes/no/maybe. You collect nominal data when there are important characteristics of subjects in the sample, so you can determine if there is any bias in responses. For example, do the responses of males differ from those of females? If the answer is: "Yes" you'll need to plan to find out why the difference.

Measurements that are nominal in nature must be organized into predefined categories that are distinct and mutually exclusive, such as, male and female, republican and democrat, green and red. For this type of data, we use labels to signify the distinctive qualities of the units of measurement, descriptive characteristics or attributes. The label is used as a means of grouping responses into the same category. Given a particular set of categories, if a unit of measurement belongs in one category (e.g., football), then by definition, that unit is excluded from any other category (beach ball, soccer ball, etc.) in the set.

For analysis, it is often convenient to assign a number to represent the category. *It is important to realize that these numbers have no numerical meaning.* They are arbitrarily chosen, since you could just as easily have assigned some other label. However, once assigned, the number must remain constant for the balance of the research, or the data will be messed up. Suppose you want to determine if the gender of the subject affects his or her responses. You can assign 1 = male and 2 = female. However, the results would not be changed if instead you assigned 1 = female and 2 = male. The important aspect of these numbers is that they are merely labels and don't have a numeric value. It would be just as valid and probably less confusing to use M = male and F = female. Alphabetic labels may be used to avoid any ideas that the data might have a numeric component.

ORDINAL DATA:

Ordinal data is data that shows a rank or order, such as, first, second, third, and so on. In contrast to the numbers used for nominal data, the rankings do have specific meaning. However, only the ordinal value of the number is used. Whatever is used for the ranking itself is nominal as it is a label. While the rankings are usually expressed in numeric form, the numbers do not represent a quantitative measurement. We may know that one item comes first and another second, but we do not know anything about the size of the interval between them. And it is possible that the time interval between first and second is different from the time interval between second and third. When a runner is the second person to cross the finish line, we only know his position relative to the previous person to cross and the ones following him. We do not know how long he took or what his time was relative to the first and/or the third runner.

The purpose of this type of data is to compare the rankings of the respondents. People can be ranked according to the results of an examination or a contest. Alternatively, you may ask people to rank specific items, such as candy, in the order of their preference. (Chocolate usually comes in a strong first!)

Examples of ordinal data include: the order in which people in a marathon cross the finish line; the ranking of people by their exam scores; the ranking of items in order of their importance or value. Unlike the numbers assigned to nominal data, the ordinal number is meaningful, since it is not arbitrary, but shows a relationship between ranked items. It is not necessary to use numbers for the rank, letters of the alphabet, since they imply a specific order would serve equally well.

Note that ordinal data naturally uses nominal data as part of the information. In the race, the runners' names are the nominal data and the ranking is the ordinal data. Tests for both nominal and ordinal data are non–parametric.

INTERVAL DATA:

Interval data provides quantitative information, such as, time taken to perform a task, age, weight, etc. Data of this type is based on fixed units of measurement such as years, pounds, or seconds (time). In other words, the interval between any one unit and the next is a fixed amount. In a long-distance race, for example, John came in first, Joe came second two minutes later, Pete was third, five minutes after Joe. Although the time interval between the times of each runner differs, since the length of "minutes" is fixed, time is interval data.

Interval data is measured on some kind of a scale with equal intervals along the scale. The data is measured in quantities referred to as quantitative data. You can compare two sets of data when you can calculate an average for each set. You can also compare the average of a set of data with the estimated average for the population. One example is IQs; so many tests have been done, that the "average" IQ is now considered to be 100. A group of people tested for their IQ may turn out to be "above average" or "below average.

Parametric tests are used on Interval data. Interval data has two dimensions. One dimension may describe the characteristic of the information, the other shows the frequency of each characteristic. The characteristic may be given a number, such as the people in a classroom with a number assigned to distinguish each member of the class: 1, 2, 3, 4...; or an alphabetic letter, A, B, C, D... Or it may be divided into equal sized intervals, such as 0-9, 10-19, 20-29, etc., with the result that the data is "grouped" into the appropriate interval. If the interval refers, for example to the ages of those being studied, 0-9 includes all children less than ten years of age, 10-19 includes all who are at least ten, but are not yet 20, and so on.

You usually obtain interval data when you wish to compare the relationship between each member of your sample and the average for the characteristic(s) you are interested in. Note that the labels of "years" or "pounds" or "distance" or "time" are nominal in nature. You can have comparisons which would provide ordinal data, such as: Data was collected for the years 1945 through 2030. The greatest number of people still living in 2030 were born in 2001, the next greatest number were born in 1995. The next was 2005 (etc.) The number of those born in 1945 were the fewest still living. The ordinal value is the *order* in which the various years are placed. For example, 2001 came first with the greatest number still living, 1995 came second, 2005 came third and so on. The Interval Data is the number of people born in each of the years being considered, however the emphasis is on the ordinal values, the ranking of years, not the interval values. Note that collecting interval data naturally also includes gathering rational and nominal data.

RATIONAL DATA:

Rational data shows a ratio; for example, suppose a freshman class had a ratio, males to females, of 3 to 5; the ratio value is written as 3:5. This means that for every 5 females, there are 3 males. If told that there are 24 students in the class you could assume that there are about 15 females and 9 males. Rational data may have an absolute zero (meaning nothing or no value) such as the Kelvin scale of temperature where zero represents the very lowest possible temperatures in the universe. A Bank Savings Account has an absolute zero: you can't withdraw more money than is in the account (not if the teller is alert).

Zero may have an actual or a relative value. If zero has a relative value such as the zero on the real number line, with positive (+ve) numbers to the right of zero and negative (-ve) numbers to the left, then this type of zero is part of the interval data set. Except in the physical sciences, it is rare to obtain rational data which, therefore, falls outside the scope of this book.

IMPLICATIONS FOR GATHERING THESE TYPES OF DATA:

When dealing with interval data, you can naturally collect the other types of data, nominal and ordinal. This means that nominal, ordinal, and interval data can all be classified, so all three types can provide nominal data. Ordinal and interval can both be ranked, but only interval type data provides quantitative information. To illustrate the differences between each type of data:

COLLECTING NOMINAL DATA (CATEGORIES):

Suppose you have been authorized to purchase new books for your library. The head librarian wants you to purchase books in the following categories with Dewey classifications: religion (200's); science (500's); literature (800's); and history (900's). The Dewey classifications of the books you intend to purchase are categories (the original assignment of Dewey numbers to the categories was arbitrary). Therefore the data are nominal in nature.

COLLECTING ORDINAL DATA (ORDERED):

You have listed all the books in each category that you would like to get. You realize that price variations may make it impossible to get them all, so you and two of your most experienced assistants independently go through each category list and assign to each book a priority of 1 (essential), 2 (useful), or 3 (nice to have). Then the three of you get together and make sure you are all in agreement concerning these priorities. The priorities produce ordinal data, showing degree of difference (1 through 3).

COLLECTING INTERVAL DATA (NUMERICAL OR QUANTITATIVE):

The budget allows $10,000 to be spent. Your library's collection is weak in the area of literature (nominal category), so you decide to give $5,000 to this category. You assign the rest of the money according to the number of essential or priority 1 (ordinal rank) books selected in each category. This results in allowing $2,400 for science, $1,600 for history, and $1,000 for religion. The dollar amounts involved here are interval data, being quantitative in nature. A dollar in one category is equal to a dollar in another category. The amount spent in one category can be added to the amounts spent in other categories to reach a total for all categories. You can even get an average of the amount of money spent on the lesser categories (i.e., excluding the literature expenditure) by adding the three amounts: 2,400 + 1,600 + 1,000 = 5,000. Then divide by 3 to get the average amount spent in each of these categories, that is, 1,666.67).

WHAT CATEGORIES OF INFORMATION CAN WE COLLECT?

When collecting information for a research study, it is probable that data in several categories will be collected. We may collect nominal by getting demographic information so we can classify the units of measurement of our study. This might include the categories of age, gender, level of education, place of birth, and so on.

We may wish to learn about the respondents' preferences or biases, to better analyze the data with reference to the intent of our study. Perhaps we collect ordinal data by asking the subjects to rank certain items, such as work procedures, reading preferences, etc. We collect interval (numerical) data, such as the cost of certain items in the budget, or the scores obtained in a test. If it seems appropriate, we may even gather anecdotal information, to determine the subjects' perception of how they felt, for example, about three different methods of instruction they received during the study.

When we gather data, we are looking for similarities and differences. We may be comparing the responses of males versus females. Or we may be comparing the effects of several methods of presenting information. We may be comparing the subjects of our study with what is "normal" or "expected" for the population. What data we collect will be determined essentially by our research question.

Some data is gathered through a questionnaire, usually to determine preferences or opinions. The data would be nominal if we were comparing responses by gender or age. The preferences could be as a ranking (ordinal) or an actual preference or opinion in a range of scores which would be interval data.

EVALUATING THE INFORMATION WE COLLECT:

A visual analysis of the information we collect is not enough to tell us whether the differences we observe are truly significant. Suppose the average length of an adult flea is three millimeters. If we measured a group of fleas and found their average length was four millimeters, we might think that as one millimeter is such a tiny measurement the difference of one millimeter is not significant. On the other hand, if we compare the average height of pygmies with the average height of "normal-sized" human beings, we might conclude that a difference of two feet is significant. Yet, an in-depth analysis may prove us wrong in both cases; and what about the effect of measuring the shortest member of the University expedition against the tallest of the adult pygmies and the tallest member of the expedition against the shortest adult pygmies. That would bias or skew the data, making any conclusion difficult to uphold. It is not only essential to test our data but also to make sure we use the right tests. We also need to consider how the data relates to the question(s) we are asking.

UNIT 1C Assignment:

1. **Given the sets of data described below, explain how you can tell if the data is nominal, ordinal, or interval.**

   a. Compare a student's mid-term and final scores.

   b. Compare male and female responses to an opinion poll.

   c. Compare the rankings of drivers in the Indy 500.

   d. Find the average time of the first five people to cross the line in the New York Marathon.

   e. Determine which kind of diet on three different farms produces the best flavored beef.

   f. People who donated blood at the local Red Cross were assessed by blood type and attitude toward donating blood.

2. Given brief descriptions of several research questions, determine which type or types of data (nominal, ordinal, interval) will likely be collected for each question.

   a. It was hypothesized that the acid content of paper in books published by Holt, Rinehart & Winston would be 0.6. A sample of 100 books published by the company during the current year was tested for acid content.

   b. It was hypothesized that a course in remedial mathematics would improve the math skills of a group of 120 students who had done poorly in math prior to attending college. A comparison was made between the 120 registrants in Mathematics 101 at the beginning of the course and again at the end of the course.

   c. It was hypothesized that there would be no difference in the number of books checked out of the university library by nuclear engineering students and psychology students. A comparison was made between 100 seniors majoring in nuclear engineering, and 100 seniors majoring in Psychology on the average number of books checked out by each group.

   d. It was hypothesized that the mean number of students using the university bookstore between 2.00 p.m. and 8.00 p.m. would be 105. Data was obtained on the actual number of students using the bookstore during those hours.

   e. The manager of the university bookstore wanted to schedule his workers so there would be an adequate number of assistants available each day of the weak and each of the hours the bookstore was open. Data was collected during each hour of the day and during the six days of the week when the store was open.

   a. Write a report of no more than two double-spaced pages, addressing the following: Explain and justify which one or more of the described types of data should be collected for *your* problem/question. Don't forget your name!

3. **Critique the reports of the discussion group, then participate in a group discussion about the answers to items 1 and 2, and discuss each of your reports. Revise your report and hand in the critiques of your paper, your original and your revised report.**

UNIT 1C Assignment Feedback:

1. Given the sets of data described below, explain how you can tell if the data is nominal, ordinal, or interval.

a. Compare a student's mid-term and final scores.

Interval data (only one person's exam scores)

b. Compare male and female responses to an opinion poll.

Nominal Data (gender) ***+ Interval Data (for the opinions which provide quantitative information)***

c. Compare the rankings of drivers in the Indy 500.

Ordinal data (the rankings) + Nominal (names of the drivers)

d. Find the average time of the first five people to cross the line in the New York Marathon.

Interval + Ordinal to identify the ranks of the five people

e. A researcher wanted to determine if the place where people lived in Utah County made a difference to those people's political affiliation.

Nominal (location) + Nominal (political affiliation

f. Determine which kind of diet on three different farms produces the best flavored beef.

Nominal + either ordinal (rank the beef flavors) or interval (on a 5 or 10 point scale) or both depending on how the questions are phrased

g. People who donated blood at the local Red Cross were assessed by blood type and attitude toward donating blood

Nominal (blood type) + interval (attitude on a five-point scale); may include ordinal if rankings are used

2. Given brief descriptions of several research questions, determine which type or types of data (nominal, ordinal, interval) will likely be collected for each question.

a. It was hypothesized that the acid content of paper in books published by Holt, Rinehart & Winston would be 0.6. A sample of 100 books published by this company during the current year was tested.

Interval; Nominal not needed to identify the books as there is no intent to distinguish between the books but only the acid content of the paper

b. It was hypothesized that a course in remedial mathematics would improve the math skills of a group of 120 students who had done poorly in math. A comparison was made between the 120 registrants in Mathematics 101 at the beginning of the course and again at the end of the course.

Interval (exam scores) + Nominal to identify the different exams

c. It was hypothesized that there would be no difference in the number of books checked out of the university library by nuclear engineering students and psychology students. A comparison was made between 100 seniors majoring in nuclear engineering, and 100 seniors majoring in Psychology on the average number of books checked out of the university library by each group.

Nominal to identify the group + Interval data for the numbers checked out

d. It was hypothesized that the mean number of students using the university bookstore between 2.00 p.m. & 8.00 p.m. would be 105. Data was obtained on the number of students using the bookstore during each of those hours.

Interval (total number of people using the bookstore) Nominal since the intent was to also distinguish between each hour (in order to get an average between those hours)

e. The manager of the university bookstore wanted to schedule his workers so that there would be an adequate number of assistants available each day of the week and each hour the bookstore was open. Data was collected during each hour of the day and during the six days of the week when the store was open.

Nominal data to distinguish hours of the day and days of the week; interval data for the number of people using the store on both an hourly and daily basis

UNIT 1D: THE "HOW TO" OF RESEARCH

PURPOSE:

One assumption made in this course is that most of you will do research with human beings, whether children, teens, or adults or a cross-section of either two or all three. But even if your research is with other units of measurement such as animals, plants, snakes, or insects, or even taste (yes, it is possible to do research on people's attitude toward the flavor of edible products!), the same principles discussed here will still apply.

The purpose of this section is to discuss the How's, that is, the method you plan to use as you do your research. Consider the information about the different types of research and decide what method or combination of methods will best enable you to complete your research study. Your final research report will require an ***Introduction*** that will describe the method(s) you plan to use. The end product of this unit is a first draft of the Introduction to your research project. A group discussion will provide an opportunity to improve your Introduction and assist others with a critique so they can improve their Introduction.

OBJECTIVES:

- Describe the elements of research that help define your methodology for any research you may undertake:
    - Historical Research.
    - Experimental Method.
    - Descriptive Survey.
    - Analytical Survey.
    - Demographic Analysis.
- Based on **your** problem statement, describe the basic elements required in the write-up of the methodology section (a.k.a. your Introduction) for **your** research study (not more than one double-spaced page).
- Based on your problem statement **and** your Introduction, determine which types of data (nominal, ordinal, and/or interval) will be collected (your introduction and problem statement may need to be adapted so you can include all three).
- Participate in a group discussion to critique each student's report and use their critiques to rework and improve your own.

INTRODUCTION:

Your research study will require an introduction which should be about a half-page description of the methodology you will use (one to two paragraphs). Answering the "how" tells the reader of your paper how you plan to gather your data, what size group you will work with and the demographics of the units of measurement. This might be described as your plan. Later in the course, you will use your Introduction to guide you in writing up the actual research you plan do and describe how you gathered your data and how it was analyzed. If you plans changed during your research as sometimes happens, you will need to add that information to your introduction, which should be a brief summary of your method(s), the body of your research paper will go into details. A helpful article, *Research Methods* can be found at the URL below. Review the 6-page article from "WRITING IT UP" to "CONCLUSION" [only the six pages indicated are needed for this unit's assignment].

http://linguistics.byu.edu/faculty/henrichsenl/ResearchMethods/RM_3_01.html

APPROACHES TO RESEARCH:

1. **HISTORICAL:** generally reviewing all research on a particular area in any field and describing, analyzing, sometimes comparing, what has been done in a particular field.

2. **DESCRIPTIVE:** organizing, describing and/or classifying without expressing judgment.

3. **OBSERVATIONAL:** = closely observing or monitoring with the ability to notice and describe significant details.

4. **EMPIRICAL:** based on, concerned with, or verifiable by observation or experience rather than theory or pure logic.

5. **ANALYTICAL:** based on theory and using logical reasoning.

DEFINING BASIC RESEARCH METHODOLOGIES:

HISTORICAL METHOD.

This type of research concentrates on what has been done in the past, with a corollary about how that research impacts on the present and the future. Many researchers have chosen to do an in-depth review of the literature for the particular field they are interested in. They complete some or all of the following: summarize the findings, compare and contrast hypotheses and study methods, critique each study on the basis of its thesis, the author's conclusions, and the validity or otherwise of the results. While they give a broad overview of the entire research on the topic, they also analyze the articles and give insight to the effectiveness of what was done, as well as pointers for future research. These researchers are more concerned about documenting all the research done in whatever field of interest they have. This results in a large document, as usually there have been a lot of studies completed in most fields. It is useful if one can identify such research in one's own field. One advantage of this type of research study is that it could shorten your search for a specific question or problem.

Another type of historical research is one that goes into ***archival records.*** An archive contains historic documents; one such archive is the Library of Congress. Some researchers search the archives for information on some historical person or event (e.g., battles in the Civil war) and their research is a treatise on that person or that event. Hopefully new information and/or insights about our history will be made available as a result of the study.

1. ***Experimental Method.***

The researcher does not simply try to answer the question, but tries out experimental procedures relating to the hypothesis. This usually entails having several groups (no crossovers of subjects between groups) who are subjected to some experimental treatment, for example, studying different methods of providing a particular topic of instruction.

2. ***The Descriptive Survey Method.***

The subjects of the study are surveyed to determine characteristics and demographics of the population; the Census is one example of a survey used to characterize the population of the whole country. Survey instruments may be used in interviews (phone, or in person with single individuals or groups), questionnaires, or observations in a controlled environment.

3. ***The Analytical Survey Method.***

The primary purpose of the research design is to compare subgroups of the population being considered. Each subgroup must be unique with no member from any other subgroup. The whole point of the study is to determine if the subgroups are the same or not. If not, to determine what causes any differences.

4. ***Demographic Analysis.***

Researchers are often interested in the demographics of a population. This is similar to the Descriptive Survey method. Demographics are categories, qualities, or characteristics of the population one is interested in. The 10-year Census is interested in the demography of the population of the United States. Many other countries also have censuses every ten years for the same reasons. The Oxford Dictionary defines demography as "the study of the structure of human populations using statistics relating to births, deaths, wealth, disease, etc."

The usual demographics sought in research are typically age, sex, height, weight, ethnic origin, place of birth, etc., depending on what characteristic the researcher is interested in. Demographic analysis is concerned with these characteristics and the analysis concentrates on the demographic information collected. The information collected is nominal data and non-parametric tests are used to analyze this data. With the possible exception of the Census, seldom does a study only collect demographic data.

DESCRIBING *YOUR* METHODOLOGY:

In your final paper, there will be two sections that deal with methodology. The ***Introduction*** must include a description of the methodology you plan to use (how you will do the research, define the terminology unique to your research area, how you will collect and analyze the data, what problems you anticipate, etc.). The ***method section*** of the documentation of your study is a description of what you actually did and how you achieved the results. The two parts should be comparable.

UNIT 1D Assignment:

1. **Read the 6 page article, beginning with *"WRITE IT UP,"* found at the URL:**

   http://linguistics.byu.edu/faculty/henrichsenl/ResearchMethods/RM_3_01.html

2. **From the article, describe what you feel will be most helpful in writing up your methodology section, emphasizing the reasons for including a methodology section.**

3. **Write the Introduction to your research using no more than one or two double-spaced pages (be brief but understandable). Base the introduction on the following:**

   a. Include a statement of your problem, your justification for studying that problem, and what you aim to achieve.

   b. Define the terminology that may not be familiar to your potential audience.

   c. Define, explain and justify the use of your chosen methodology for your research (briefly include how you plan to collect the data, what you expect the results to be, and how it applies to your problem statement).

   d. Discuss how the problem statement, the methodology, and the sample size interact.

   e. Provide copies for the rest of your group (with your name) and collect a copy of each of their introductions.

4. **Review the reports of the rest of the group, analyze them and write a critique with suggestions for improvement with reference to their description of their methodology and its application to their problem statement. (Be prepared to justify your suggestions).**

5. **Participate in a group discussion to critique each student's report, then use their critiques to revise your report.**

6. **Hand in the critiques received, your original Introduction, and the revised Introduction.**

UNIT 1E: STAGES IN THE RESEARCH PROCESS.

PURPOSE:

There are certain steps that should guide your research. This unit describes in detail these steps which also provide an outline for what should be included in your final report.

OBJECTIVES:

- Define and/or describe all aspects of doing basic research.
- Identify a particular research question and form at least one hypothesis.
- Prepare to perform a literature search in the field of your research question [UNIT 2A].

THE RESEARCH PROCESS:

There are certain major steps in the research process. These steps are discussed here to provide a framework for the rest of your research activities. You will get the opportunity to practice each step. As part of the process, you will review and write critiques of other class member's work and take part in group discussions so that everyone can improve the quality of their final product.

1. **DETERMINE WHAT QUESTION IN YOUR FIELD YOU WOULD LIKE TO STUDY AND WRITE IT AS A PROBLEM STATEMENT:**

   a. Decide what field, area or domain within your major interest you would like to research for a small study.

   b. Properly define a research problem in the field, area or domain you have chosen and write it as a **Problem Statement.** (What do you want to know and/or find out?)

   c. Justify your research by explaining why it is of value to yourself and others in your field. (Why do you want to research this question?)

   d. Examine the field of your proposed study and your problem statement to identify and list any terminology that you will need to define.

2. **DESCRIBE THE METHODOLOGY YOU WILL USE:**

   a. From the various methods, described in Unit 1D, for doing research, select those you feel are of most value to use in studying your research problem.

   b. Prepare the ***Introduction to your research report*** by describing what methods of research you plan to use.

3. **DEVELOP PRELIMINARY QUESTIONS TO PRIME YOUR LITERATURE SEARCH:**

   a. What have others said or done in the area in question?

   b. What has happened in the past?

   c. Survey what's happening now, how do people respond?

   d. What parameters (a list of important characteristics or **descriptors** suggested by your problem) do you need to specify in order to complete 3. the literature search.

4. **LITERATURE SEARCH:**

   a. Ask your faculty advisor for a list of journals published in your discipline and check the University Library to find out where they are located.

   b. Use ERIC (an Internet site) to find out what specific professional articles are available on ERIC.

   c. Make an appointment with a librarian for assistance to search the journals in your university Library (there may be indexes to the journals.

   d. Analyze each article in terms of its thesis, supporting evidence and suggestions for further research.

   e. Determine if the researcher(s) missed some important aspects of the research, record your findings and their importance. Assume you are providing the "hindsight" the original researcher may have missed.

5. *The Product Of Your Search Should Be A List Of Articles Dealing With Your Problem Area:*

a. Apply this information to your problem statement and determine articles which:

b. Have implications for *your* problem statement and research or...

c. Give essential background information or ...Give details on research studies in the same field as your problem statement.

6. *Annotated and Simple Bibliographies:*

a. Use the list to create an annotated bibliography using your department's bibliographic style, ***making sure the original source of the articles in the bibliography is clear***.

b. The purpose of the annotated bibliography is to add explanatory notes and/or a brief description of the salient features of each article with comments about your perceptions of each article. This will be a valuable tool in the future. If you ever needed to access one of these articles again, you would know what each article was about.

c. List only the articles that are pertinent to **your** study in the form of a simple bibliography.

d. Selected quotes may be useful in supporting your conclusions about the literature found but ***make sure you include the source of your quotes***.

7. *Theory development (where little or no research has been done):*

a. This step may be a necessary part of planning larger research studies. The small study, you will complete this semester, may be regarded as your personal theory development. You may choose to complete another study which, though it may be more than just a duplicate, could be regarded as fulfilling item **d. and e.** below.

b. Set up an experiment.

c. Gather and analyze data.

d. Form conclusions.

e. Suggest further research possibilities.

f. Duplicate the experiment with a different sample of subjects.

8. *Complete the Introduction:*

a. Based on the results of your literature search and your analysis of the articles found, refine your introduction to your research study which should describe:

b. What research has already been done in the area of the reported study.

c. What conclusions were made by the researchers who reported the studies described.

d. Any questions from their research that either the author(s) describe or which you identified and feel you can answer with further studies.

e. **NOTE:** ***The introduction is your opportunity to justify the research you have planned and can lead to defining or redefining your hypothesis.***

f. Describe your methodology (how you plan to collect the data, what you expect the results to be and define any terminology specific to your research study).

9. WRITE HYPOTHESES:

a. Convert your problem statement into an hypothesis.

b. Describe the data you need to collect to test your hypothesis.

c. Determine how you are going to collect the needed data.

d. Describe what you expect to result from your research.

10. CREATE THE TOOLS FOR THE STUDY:

a. Collect and create any apparatus, equipment, survey instruments and/or questionnaires needed by the units of measurement of your study in order to provide the data.

b. If you plan to use interviewing techniques, plan the questions to be asked and the method. (Are you to do the interviews or will you arrange for someone else to do them. If someone else how will you train them? How will you make sure they stick to your guidelines?)

11. WRITE A PROPOSAL FOR THIS RESEARCH:

a. Use all you have written to this point to write a proposal as if you were asking for funding for the study.

b. After the discussion group works on the proposals, refine yours.

c. You will be required to submit it to your Faculty Advisor and your department Chair. [Make sure the two people are aware that this is an assignment and ask them to critique your document.]

12. DATA COLLECTION & ANALYSIS:

a. Collect and analyze your data; for larger studies you will need to work with a statistics graduate assistant to use the computer program ANOVA for the analysis.

b. However this textbook will only briefly review certain aspects of ANOVA.

13. CONCLUSIONS:

a. Study your data and analyze the results carefully to determine what conclusions you can come to.

b. Show how your hypothesis is or is not supported.

c. Discuss the significance of the conclusions and discuss what else you could have done in data gathering (for a thesis or dissertation, it may be necessary to repeat your study gathering the new data).

d. Develop questions, if any, for follow-up research.

14. FINAL REPORT:

a. Write, in the form of a professional article, a description of the research done, from the first step through to the final step.

b. This report will be the final assignment for this semester.

UNIT 1E Assignment:

There is no specific assignment for this unit. It is probable that you will need to continue to review the assignment from 1D. The following questions are here for review:

1. **Write the Introduction to your research using no more than one or two double-spaced pages (be brief but understandable). Base the introduction on the following:**

   a. Include a statement of your problem, your justification for studying that problem, and what you aim to achieve.

   b. Define the terminology that may not be familiar to your potential audience.

   c. Define, explain and justify the use of your chosen methodology for your research (briefly include how you plan to collect the data, what you expect the results to be, and how it applies to your problem statement).

   d. Discuss how the problem statement, the methodology, and the sample size interact.

   e. Provide copies for the rest of your group and collect a copy of each of their introductions.

2. **Review the reports of the rest of the group, analyze them and write a critique with suggestions for improvement with reference to the their description of their methodology and its application to their problem statement. (Be prepared to justify your suggestions).**

3. **Participate in a group discussion to critique each student's report, then use their critiques to revise your report.**

4. **Hand in the critiques received, your original Introduction, and the revised Introduction.**

SECTION 2: BASIC RESEARCH PROCESSES.

PURPOSE:

When you are doing research, you cannot do it in a vacuum. You must set your research in the global background of your specific field. Since part of the intent of your research is to add to the information currently available for researchers in your field of study, you have to know what has been done before.

You have already decided the main thrust of your current research study in the form of a problem statement. The only way to determine how the concept in that statement fits in to your field is to do a ***literature search***. In the past, researchers have examined the professional journals, many of which have been published for quite a number of years. The search task for a beginner would mean having to search what may amount to years of publications simply to find a single study that covers your problem statement area. Since one of the requirements in reporting research is to provide a bibliography of related studies, the task is a little easier once you have identified that first study. But the bibliography may refer to publications from years ago and the researcher may have to spend many hours in the university library.

Today the search task has been simplified because the Internet allows you to search for such publications. The literature search is made easier partly because of "ERIC," an acronym for "Educational Resources Information Center," a site that has collected many articles and other publications in the educational fields. Journals published since 1966 have been fully indexed. There are also a number of non-journal articles, papers, theses, dissertations, and other materials available through ERIC.

Once you have an idea of what has been published about your problem area, you can complete other processes in your research study: converting your problem statement to a research hypothesis, develop a process for collecting data, and writing a research proposal. Unit 2 teaches you how to do all that.

OBJECTIVES:

- Use ERIC to find articles and other literature to provide support for you area of research.
- Convert your problem statement to a research hypothesis.
- Determine what data you will need to test your hypothesis.
- Write a research proposal, to be reviewed by your faculty advisor and you department chairman.

INTRODUCTION:

One aspect of research that is often missed or short-changed is the professional literature. A research study is supposed to add to the knowledge in the field in question so you cannot effectively start until you know what else has been done in the area of your interest.

THE RESEARCH PROPOSAL:

The basic purpose of a research proposal is to obtain funding for your research. Most of the larger studies need funding, particularly for the development of data collection instruments, such as surveys and questionnaires, payment for persons who will collect the data, analyze it, and perform other necessary tasks.

Even though you will not need funding for this semester's research, future research will. In fact, most of the larger educational research studies require funding, sometimes in the thousand or even millions of dollars. There is much wisdom in giving you experience in creating a research proposal. After gaining understanding of the various aspects of doing research, you will in UNIT 2D write up a proposal. Only your class instructor will actually grade your proposal, but you are required to present your proposal to your faculty advisor ***and*** your department chairman. Make arrangements beforehand with these professors, explaining that what you need is for them to review the paper and determine if it would be "good enough" to get funding. Any suggestions from either would be of real value.

UNIT 2A-1: THE LITERATURE SEARCH.

PURPOSE:

At this time, you should have written a problem statement suitable for a research study and have also decided on the size of sample and the data that will be the focus of your study. Note that you are assigned to bring in samples of each type of data so you can gain experience with all types. You should have also written an introduction to your study describing your anticipated research method. The next step is to do a literature search in your subject area. It is important to analyze the literature you find. The analysis may reveal a need to revise both your problem statement and your introduction.

As you will probably do further research in the area of your graduate major, you should create an annotated bibliography of the articles found, then, as part of the requirements for this unit, select 2-4 articles most applicable to your problem statement. You need to begin an annotated bibliography as a basis for future studies. "Annotated" means that you attach to the title of each article explanatory notes, critical analyses, commentary on relevance, etc., to provide a valuable tool when you are preparing to do further research. A simple bibliography is a list of articles that are relevant to your current research.

OBJECTIVES:

- Analyze your own focus (theme) and scope of your research study using your statement of the problem and determine if there are any sub-problems. If there are, write a problem statement for each sub-problem.
- Using your problem statement and introduction as focus, search professional literature for those articles that relate to your research.
- Evaluate the literature you find in terms of its thesis (problem statement or theory), supporting evidence, and suggestions for further research.
- Write an annotated bibliography of **all** articles using abstracts (written summaries) of each article; use the style required by your department.
- Identify two to four of the articles found that most closely tie in with your research problem and write a simple bibliography of these articles using the style required by your department:
    - Examine your statement of the problem in light of the literature identified in this objective, narrow or expand your focus, then explain how the selected literature has changed or strengthened the focus of your research question/problem statement.
    - Compare and briefly describe the methodology, the evidence they present, and their conclusions, then give a reasoned critique of each study. If their methodology has suggested other methods to use in resolving *your* research problem, explain what and why.
    - Analyze the articles to answer the following: What questions do the articles suggest for further research and has the researcher "missed" any questions that you feel could have been included in their study.

INTRODUCTION:

When doing research, the first step is to decide what area in your field of interest you would like to research (you can change or adapt yours as you complete this unit). You already have a statement of the problem and an introduction describing your anticipated methodology. One aspect of research that is often missed or short-changed is the professional literature. A research study is supposed to add to the knowledge in the field in question so you cannot effectively start until you know what else has been done in the area of your interest. Carry your Research Notebook which should include your research plan and its execution plus any revisions you may make as you work on your study).

You have already been directed to a helpful article, *Research Methods;* the third page discusses the literature review; found at the URL

http://linguistics.byu.edu/faculty/henrichsenl/ResearchMethods/RM_3_01.html

THE LITERATURE SEARCH PROCESS:

1. Begin with your question.

   a. Look at the area that interests you, and ask questions: Why does such-and-such appear to be so? What is the cause of this? What will be the effect of that? What does it all mean? Do I have a possible explanation (theory) about why it's like that?

   b. Use your problem statement to refine the area you would like to research (what and why).

2. Analyze the problem thoroughly.

   a. Define all possible variables of your problem.

   b. Describe which parameters you feel would best fit the problem to help focus your literature search.

   c. Analyze your problem statement and any identified supporting concepts for *key words* to use in the literature search.

3. Find out what has already been done in the area.

   a. This is your literature search - the library has computer programs (ERIC) that enable you to find articles that relate to **your** problem statement. **[Refer to Handout #1: Using the Internet to search ERIC.]**

   b. Finding out what has already been done helps avoid unnecessary duplication of effort.

   c. Analyzing what has been done also helps you to define the limits or parameters of **your** problem.

   d. Select two to four articles that are of most value in tightening the focus of your problem.

4. Analyze each selected article.

  a. Determine and write a critique of the problem statement in each author's research.

  b. Did the author identify **all** the variables in the problem? If not, what other variables would you add?

  c. QUESTION EVERYTHING . . . Analyze and critique the research articles with the following questions in mind:

  d. What method of collecting data was used? Did the author give a clear description of their methodology?

   - Was anything left out or could be added to make the study more helpful?
   - What assumptions (correct and/or incorrect) were made?
   - Were the author's conclusions valid (especially in view of what was left out or incorrect?
   - What was well done and what could have been done more effectively?
   - How useful is each article to your understanding of how you can conduct your research and refine your question?
   - **Use your answers to help with your annotated bibliography.**

5. Identify Research conclusions.

  a. Were the conclusions of the author(s) clear or did you have difficulty understanding how they came to those conclusions?

  b. Did the conclusions answer the author's problem statement and/or hypotheses? What would you change?

  c. What was valid or invalid about their conclusions?

  d. Write a critique of the author's conclusions.

6. Questions for further study.

  a. Identify any questions or suggestions in the selected articles that give possibilities for further research.

  b. Identify any other questions for further research that could have been added to each article.

7. Create an annotated bibliography of all articles found (make sure you use the bibliographic style required by your department). NOTE: Annotated means explanatory notes which, when you later refer to them, will assist you in remembering what the article was about; the author's abstract or summary will assist you in completing this requirement.]

  a. List each article you found with a one paragraph description of the research. Write enough so that if you needed to refer to any of these articles in the future, you would know what it was about.

  b. Extract those articles that will be useful for your study, reviewing your problem statement(s) and use the knowledge gained from these articles to refine your problem statement(s) if necessary.

  c. This is required for your current literature search.

  d. For future research an annotated bibliography is RECOMMENDED. Many times people have a recollection of a research study that they need but cannot remember its title or where to find it. With an annotated bibliography in your research notebook or on your computer, you will be able to find what you need far more easily.

UNIT 2A-2: SEARCHING ERIC FOR ARTICLES

(ERIC: Educational Resources Information Center)

PURPOSE:

For many years, people have been working on Thesis and or Dissertations in Graduate School. In addition, faculty members have been publishing articles in the publications relating to the field in which they teach. During that time, an entity called ERIC, an acronym for "***Educational Resources Information Center,***" has come into existence. The main object of ERIC is to provide a resource of educational articles for these authors. This section provides information on how to access ERIC.

OBJECTIVES:

- Describe the major segments of the Educational Resources Information Center, including the articles on the ERIC Copyright Policy, the ERIC Collection, and the ERIC Site Map.
- Describe how to access ERIC articles through a Basic Search and an Advanced Search.
- Describe how to use the ERIC Thesaurus to aid in your search.
- Search ERIC for articles in the area of your problem statement.
- Write an annotated bibliography for the articles found in ERIC, using the bibliographic style of your department.

INTRODUCTION:

ERIC is a collection of professional literature available to assist in writing professional articles for journals and in developing research and the writing of a Thesis and or a Dissertation. To access ERIC, here is the URL for the ***Home Page*** (illustration below):

http://www.eric.ed.gov/

The home page gives basic information about ERIC. Note that there are several languages that you can use when accessing ERIC.

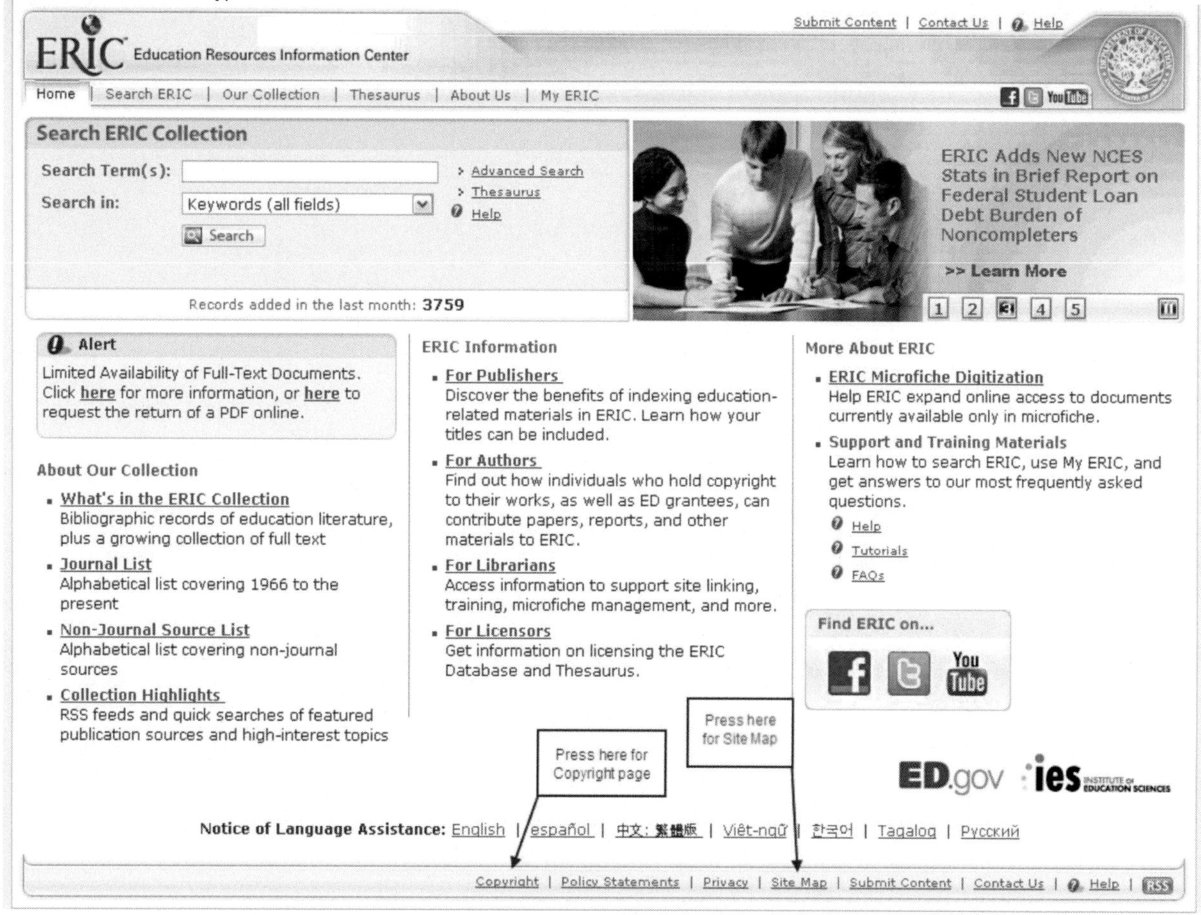

Figure 3: Follow the directions in the two boxed items.

Read the copyright information. This copyright policy applies to use of a ***complete article or document*** from the ERIC website. As a researcher, you can quote an appropriate excerpt from any article in ERIC, subject to the general U.S. Copyright laws.

Return to the HOME page and press: Site Map. You can also use the following URL to take you directly to the site map:

http://www.eric.ed.gov/ERICWebPortal/resources/html/sitemap.html

At the site map: note the section at the bottom of the middle column: "MY ERIC." Click on Register to Use ERIC to create a login so you can save your searches – that way you won't have to recreate any search you've made in the past. This saves spending time memorizing, too.

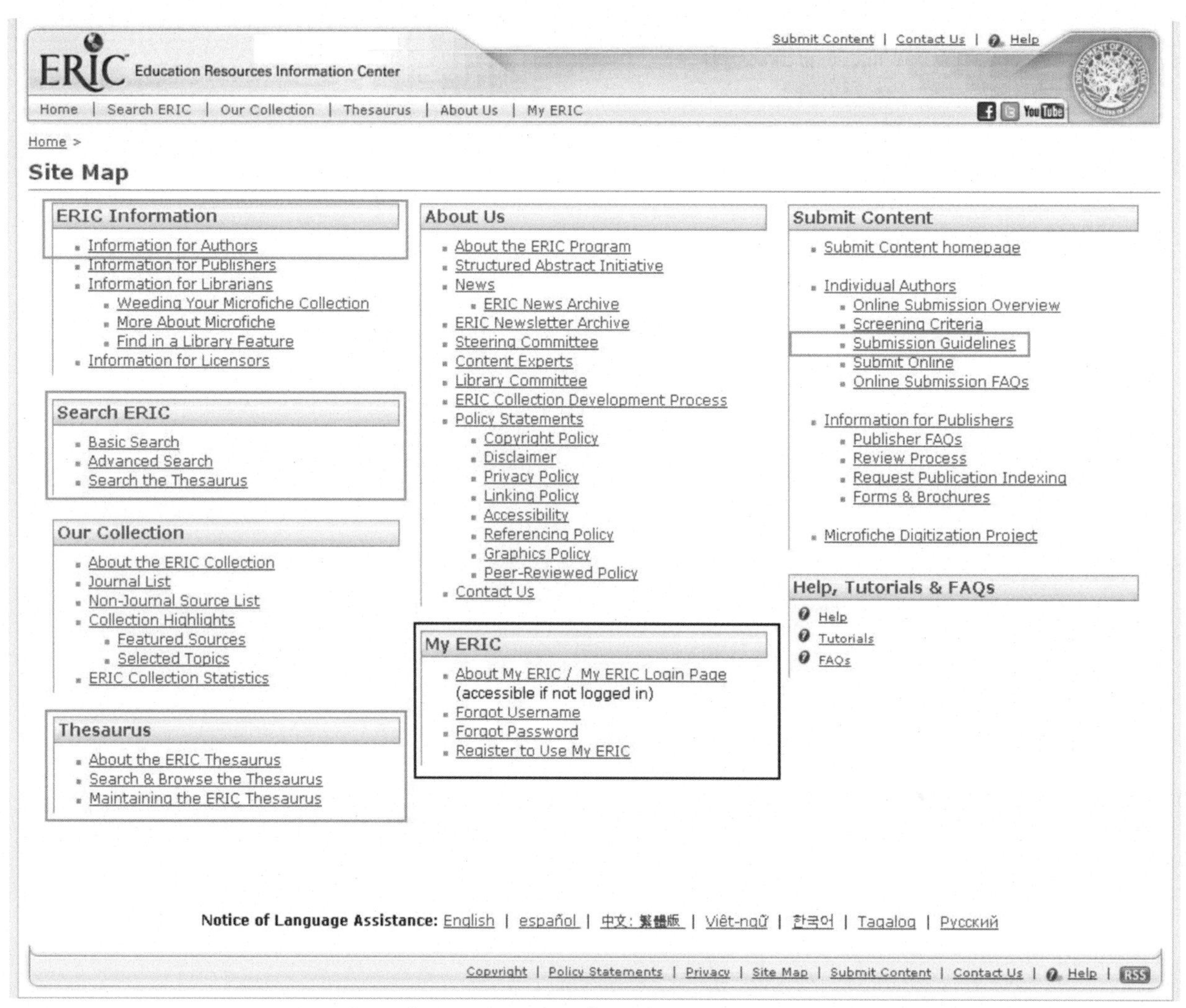

Figure 4: Important items from the ERIC Site Map.

The items that are boxed in yellow are useful in ERIC searches. As a student or faculty member who wishes to do a literature search, you are considered to be an "author." Click on the "Information for authors" and review that information. You may wish to submit the final paper for this course to ERIC. Read the information in the box on the right of the screen titled "Submission Guidelines." Follow the guideline if you do submit a paper.

The "Search ERIC" box on the left side shows that there are two level of search: Basic and Advanced. It also mentions "Search the Thesaurus." The Thesaurus box, at the bottom of column one allows you to learn more about the Thesaurus and give you access to it. When doing a search, there are two ways you can access specific information: "***Keywords***" which refer to words contained ***within*** the articles in the topic you are searching for. and "***Thesaurus***" which is a sort of dictionary of terms that apply to what you are looking for. They may or may not be within the article.

Go back to the HOME Page. Immediately below the ERIC LOGO and third along the bar is ***"Our***

Collection." Click on that to find information about what's in ERIC. About half-way down the page is: ***"View Sample ERIC Journal Record."*** Click on that which takes you to a sample record which is from one of the professional journals.

Along the left edge of the sample are descriptive notes. For example, the items in the record that are highlighted are the "ERIC Number." When you locate an interesting ERIC Record make a note of that number and include a description of what the record is about so you can go to it again if you need to. In this particular sample record, there are only two descriptors both of which come from the Thesaurus. The descriptors are intended to be subject-specific to aid in information search and retrieval. Any number of descriptors may be used. The note below the definition of descriptor refers to "Keywords" which are words you expect to find in the record iteself.

"Peer-reviewed" indicates the record has been published in a journal that ensures that peers (colleagues in the same field) have reviewed the record. This is supposed to ensure the record is more likely to be of value to other researchers. The abstract is the author's (brief) narrative description of the record.

There is a second sample of an ERIC Document as opposed to a journal record. The ERIC document at the time of submission to ERIC was not published in a professional journal. Click on the sample of an ERIC Document to review the differences and similarities. In this sample, there are more authors and the list of descriptors is longer. Note the abstract is much longer. There may be times when the abstract is too long to ptint in the record. If this happens, there will be a place to click to obtain the rest of the abstract. Note that in this sample the abstractor was "ERIC," that is, one of ERIC's employees has provided the abstract.

EDUCATIONAL LEVEL:

Near the bottom of the sample, the educational level of the article is shown as "Elementary Secondary Education." When searching for any record, you can speicfy what educational level you wish the information to relate to. You may confine it to "elementary education"; a specific class level such as "pre-school" or "kindergarten" or a specific grade number; "middle school"; or "Postsecondary"; or you could try a range of classes such as "K-12."

SEARCHES:

There are two types of searches you can make. The first is ***BASIC SEARCH*** and the second is ***ADVANCED SEARCH.*** You can also access the ***THESAURUS***.

BASIC SEARCH:

The ***"Search In"*** field provides the following choices:

- Keywords (all fields)
- Title
- Author
- Descriptors (from Thesaurus)
- ERIC #

NOTE: Every item the search finds contains one ERIC number. So if you wish to see a specific entry, select the ERIC number in the ***"Search Term(s)"*** field**.**

ADVANCED SEARCH:

You can add more rows for the descriptors if necessary. Also "AND" means both descriptors must apply to what the search finds whereas "OR" finds articles where either descriptor applies. If there is a descriptor you don't want to include in your search, you can use "NOT."

The Advanced Search allows for additional search fields. It also allows you to change the publication dates (if you know a particular article was published say between 1995 and 2000, you can insert these dates). You can also select a publication type; and the alphabetized educations levels (you could select ***Early Childhood Education*** and ***Grade 1*** and ***Kindergarten*** for example.

Near the bottom of the "***Our Collection"*** page is the information about getting started with ERIC.

EXAMPLE OF USING THE THESAURUS:

The thesaurus gives you access to the "descriptors" (key words for your search). You may have to hunt for what you need. For example, I put in the word "***Piaget"*** to find out which articles had reference to him. No luck. So I inserted the word ***conservation***, which produced the following with my notes on the right:

- Conservation (Concept). . . . . This one contains Piaget's ideas.
- Conservation (Environment)
- Conservation Education. . . . . So does this.
- Energy Conservation
- Hearing Conservation (2004)
- Soil Conservation (2004)

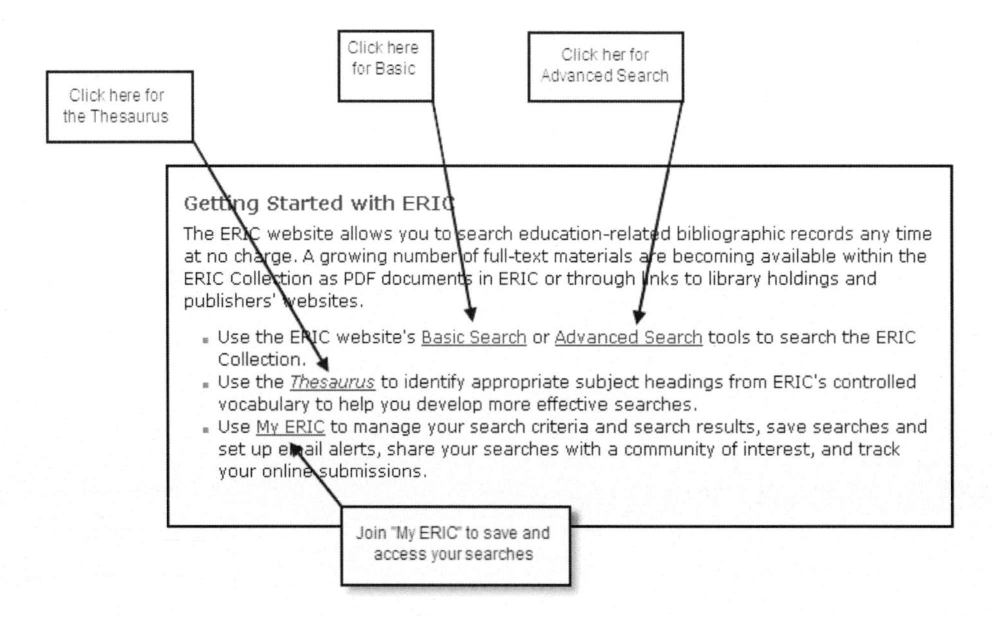

I checked both Conservation (Concept) and Conservation Education which is when I found that other types of conservation are intermingled with Piagetian Conversation. So I continued to search; in the Thesaurus search criteria, I typed ***Piagetian.*** This brought the descriptors:

- Piagetian Stages
- Piagetian Tasks
- Piagetian Theory

Selecting any one of these gave the ***Basic Search*** fields. I clicked in the box for ***Stages*** and the search program combined Stages AND Theory, which gave 229 Results at 10 items per page. I selected ***Tasks*** and again the search program combined Tasks AND Theory. This combination gave 54 results. The next choice I selected was ***Theory*** which became the only descriptor used in the search. This time there were 887 results. These results show that the outcomes you get are based on the set of descriptors you use.

MY ERIC:

If you anticipate using ERIC fairly frequently, you may want to register with MY ERIC (free) because then you can save your search results as well as do new searches.

UNIT 2A-2 Assignment:

1. Write a report of no more than four to five double-spaced pages and include the following, making sure each item is clearly identified. Prepare the report and hand them out a week in advance to share with other members of your class:

    a. Your report should begin with your **Introduction** and describe the data you plan to collect.

    b. Describe what you learned as you completed the literature search including how the search helped define or refine your problem statement/question and how doing literature searches may help you in future research studies.

2. For the selected articles, compare and briefly describe the methodology, the evidence they present, and their conclusions, then give a reasoned critique of each study (include any changes the articles may have suggested for your own research).

3. On a separate page, using the article and each article's abstract prepare a full annotated bibliography (brief summary and/or explanatory notes of each article, so that if you ever wanted to refer back to that article, you'd know what it was about and where to find it). <u>Use the required bibliographic style of your department, based on your graduate major</u>.

4. Prepare a simple bibliography of the articles you have selected as pertinent to your research focus. Briefly describe your findings from this literature that apply directly to your study and include a bibliography at the end of your report. <u>Use your department's required style</u>.

5. Review the reports of the rest of the group, analyze them and write a critique with suggestions for improvement; be prepared to justify your suggestions.

6. Participate in a group discussion to critique each student's report; then use their critiques to revise your report.

7. Hand in the critiques, your original report, and your revised report. Include a copy of each article you selected as pertinent to your study.

UNIT 2B: RESEARCH HYPOTHESES

PURPOSE:

When writing your hypothesis, you start with your problem statement. The wording of the hypothesis should be a statement which brings into focus the ***critical decision or issue*** in your research – What you believe to be true or what you would like to prove. Your statement, hypothesis and the analysis of the data you collect allows you to make the critical decision with a certain level of confidence. Most researchers select a level of confidence of 0.5 (95% confidence that your sample is a valid sample of the population), or the level of confidence may be 0.1 (99% confidence).

OBJECTIVES:

- Using the problem statement developed in Unit 1, write an hypothesis to show what direction your research should take and explain why it is appropriate.
- Examine your hypothesis and write each in both positive (test hypothesis) and negative terms (null hypothesis).
- Participate in a group discussion.
- Provide a copy of your report for each student in the discussion group a week in advance.
- Write a short critique of each report you receive and be prepared to discuss your critiques and the validity of your points.
- Discuss the experience in writing hypotheses and how each hypothesis could be improved. ***[There is no "right" answer; there is a need to really think through the process so you can improve the quality of your hypotheses.]***
- Use the critiques and discussion to rework the hypothesis you develop in this unit.

IMPORTANCE OF GOOD HYPOTHESES:

When developing a research study, you must be certain that your hypothesis is clear and pertinent to your question or problem statement because that hypothesis will direct your research study. Your hypothesis will also help determine the ***population*** for your study, the ***sample size,*** and the ***type of data*** you collect. In a similar way, population, sample size and type of data help to define your hypothesis. You must be certain that the data you collect will enable you to reach valid conclusions. Note: there is a possibility that any apparent differences you find are not significant but the result of chance. You need to be aware of the possible effects of chance and probability.

UNIT 2B-1: WHAT IS A RESEARCH HYPOTHESIS?

PURPOSE:

The ***Introduction*** to your research project was the product of UNIT 1C. *Answering the How* [to do research]. It included a statement of your problem, your justification for studying that problem, and what you aim to achieve. You have done a Literature Search (Unit 2A). As part of the search, you analyzed and developed the focus (theme) and scope of your research study. You also evaluated each article in terms of its thesis (problem statement or theory), supporting evidence, and suggestions for further research. All of this will provide a foundation for developing your own hypothesis. Your hypothesis should include a statement showing what you believe to be true or what you expect to find out by the time you complete your research project.

OBJECTIVES:

- Review the literature you found as a result of your Literature Search and extract the hypotheses that were used by each researcher.
- Compare these hypotheses for common and for unique aspects with emphasis on how the hypothesis impacted each author's research.
- Identify any terminology that may be unique to your major field of study or which are open to misinterpretation; list each term with an appropriate definition.
- Based on the work you have done on your problem statement, determine what it is you **really want to find out** and, during the class discussion, develop one major hypothesis.
- Determine the population of your subjects and the size of the sample you will need for your study; the number should be small probably no more than 10-15 units of measurement. If you feel it is essential to have more than that number, be prepared to justify it.

INTRODUCTION:

The mind of any intelligent human being gathers information obtained through observation and experience. This information or ***data*** is sorted and categorized and an explanation is formed in the individual's mind. As more data is collected, the explanation is reinforced, adapted or discarded. This process seems to be inherent in the nature of humans.

HUMAN NATURE:

Based on our perception of the world, each of us makes assumptions and/or predictions. For example, after Jane had been driving an automatic vehicle for some years, she knew that all she had to do to get her car started was to insert and turn the key as far as it would go. This knowledge is based on data she'd gathered -- a large number of successful attempts to start her car. When her car failed to start, she mentally searched for a reason. For example, she may have considered one or more of the following possible explanations (hypotheses): The battery is dead... or... I'm out of gas... or... there's something wrong with the alternator... and so on. After reviewing these hypotheses, she looked for further evidence to try and eliminate the least likely explanations. In other words, she tested her hypotheses.

THE FORMATION OF INFORMAL HYPOTHESES:

Even very small children form hypotheses, although they may not have the verbal skills to express their hypotheses in words. Never-the-less, as the child gains experience, they form hypotheses to help explain their world. For example, a child of three months has just learned to roll over. In trying to roll, he falls off the bed. If he repeats this behavior several times, he will stop trying to roll over. Why? Because, although he may not actually think in these terms, he has concluded: Every time I roll over, I fall. When I stop falling, it hurts. So (hypothesis) if I roll over, it will hurt. In this case, his hypothesis is incorrect - it is the "when-I-stop-falling" that causes the hurt.

Another child, aged seven months, is watching her mother play "peek-a-boo." Mother is hiding behind a chair and pops up alternately from the left and the right side of the chair. After playing this game for a while, the child begins to anticipate her mother's moves. After she pops up from the left of the chair, she looks towards the right for her mother's next appearance. After she pops up on the right, she looks to the left. She is predicting where mother will next appear, based on prior experience. In other words she has made a (non-verbal) hypothesis. If Mother continues to alternate her appearances, the child's hypothesis is correct. Trouble is, because we are human, we may change the rules in the middle of the game!

The process of finding explanations for the patterns and trends we observe in life is similar to, though far less formal than the research process. Although we don't normally use the term "hypotheses" for our explanations, each explanation (hypothesis) is based on data we have gathered and analyzed. Whether the hypothesis is accurate will depend on the quality of the data and whether we have enough information.

THE NEXT STEP IN YOUR RESEARCH STUDY:

When performing research, the researcher follows a similar process. However, while informal hypotheses are acceptable for most activities in life, research studies require more formality in the approach. Doing research, each of us must take care to be as objective and impartial as possible.

In Units 1A through 1E, you completed the following steps of the research process:

1. Begin with a question [pre-hypothesis].
2. Make a clear statement of the problem [What do you want to find out?].
3. Analyze the problem thoroughly [define definitions and describe how you will prove or disprove your hypothesis.]
4. Find out what has already been done in the field of your question/problem statement [Literature search.]

DEFINING THE RESEARCH HYPOTHESIS:

By definition, a ***hypothesis*** is a statement about an idea, usually theoretical. The theory applies to what you may assume is happening but which you currently may not fully understand. Note the terms ***theoretical*** and ***assume***. Your hypothesis is to be derived from your question or problem statement and should relate to a theory of why you think something you have observed is happening, why it is a problem of interest or why answering the problem would provide your fellow researchers with useful information. The hypothesis could also account for any differences you are trying to establish.

Review again the professional literature you identified as relating to your problem. Examine all hypotheses and determine what theory the researcher had in formulating that hypothesis. As you review the research description of what was done, identify any assumptions the researcher made. Apply this information to searching for an explanation (theory) for how your problem can be resolved.

There is circularity as you develop hypotheses: The hypothesis allows for some form of statistical analysis. The wording of the hypothesis depends in part on the type of statistical test the researcher will make. The test is dependent on the nature of the data being collected. The nature of the data will affect the wording of the hypothesis and help determine what statistical tests you will need to do. You must be certain that the data you collect will enable you to reach valid conclusions. Note: there is a possibility that any apparent differences you find are not ***significant*** but the result of chance. You need to be aware of the possible effects of chance and probability. Realize that when using normal English, the word "significant" means "important." It is possible for someone's research to turn up something significant statistically, never-the-less, it may not be very important! So, for our purposes in research analysis, the term "significant" means ***probably true*** or ***not due to chance.***

A ***research hypothesis*** is an assumption about a population ***parameter***. This assumption may or may not be true. As mentioned before, the best way to determine whether a research hypothesis is true would be to examine the entire population. Since that is usually impractical, researchers typically examine a random sample from the population. If testing the sample data produce results consistent with the research hypothesis, the hypothesis is accepted; if not, it is rejected. We must make a clear statement of the problem because we want answers to our question. But our answers are not valid unless we can provide statistical "proof."

What we would like to prove forms the basis for our hypothesis and "proof" is based on ***statistical significance,*** that is, analysis of our data shows there are significant differences.

UNIT 2B-1 Assignment:

1. Review the literature you found as a result of your Literature Search and extract the hypothesis or hypotheses that were used by each researcher. *If you are unable to identify a researcher's hypothesis, analyze their study and write what you think their hypothesis should be. (What are they trying to prove?)*

2. Compare these hypotheses for common and for unique aspects with emphasis on how the hypothesis impacted each author's research.

3. Identify any terminology that may be unique to each study's major field or which seem open to misinterpretation adding any that the author missed; list each term with an appropriate definition. [You should already have identified terminology in your own field, but review to make certain you have all that are needed.]

4. Identify any terminology that may need to be explained by each classmate for their major field of study. Highlight any terms used by your classmates that you feel needs a definition. Explain why.

5. Based on the work you have done on your problem statement, determine what it is you really want to find out and, in conjunction with the class discussion, write in words what you think your hypothesis should be.

6. Determine the population of your subjects and the size of the sample you will need for your study; the number should be small probably no more than 10-15 units of measurement. If you feel it is essential to have more than that number, be prepared to justify it during the discussion.

UNIT 2B-2: CONVERTING PROBLEM STATEMENTS INTO HYPOTHESES.

PURPOSE:

Currently, you have only a research question and a problem statement. This unit is designed to help you begin with your problem statement and convert it into a hypothesis. If you have identified supporting concepts, each concept needs to be converted to a hypothesis as well. Ensure that the sub-problems' hypotheses are supportive of the main hypothesis. If you have too many sub-problems, you should refocus your problem statement to eliminate the extras. You will be better able to plan your current study by reducing the number of hypotheses you work with (for a small study, one hypothesis or two at the most is best).

It is recommended for this first study that you concentrate on a single hypothesis (in both a null version and in an alternative version). Also make sure your final hypothesis requires you to collect interval, nominal AND ordinal data.

OBJECTIVES:

- Define the null and alternative hypotheses, list the hypotheses in your selected articles, and identify which ones are null hypotheses.
- Using the hypothesis you developed to show what direction your research should take, write it in both negative terms (null hypothesis) and positive (alternative hypothesis).
- Discuss the two types of error that may be made when testing the null hypothesis.
- Determine what type of data: interval, ordinal, and nominal must be collected in order to prove or disprove your hypothesis. Note: Most of your data should be interval, but other data should include categories or labels, that is, nominal data, such as name, gender and other identifying characteristics of the subjects and ordinal data, such as class rankings (grades), preferences on a five point scale, or other data that can be placed in a ranked order.
- Participate in a group discussion.
    - Provide a copy of your report for each student in the discussion group a week in advance.
    - Write a short critique of each report and be prepared to discuss your critiques and the validity of your points.
    - Discuss the experience in writing hypotheses and how each hypothesis could be improved. ***[There is no "right" answer; there is a need to really think through the process so you can improve the quality of your hypotheses.]***
    - Use the critiques and discussion to rework your hypothesis.

INTRODUCTION:

By definition, a ***hypothesis*** is a statement about an idea, usually a theoretical idea, applying to what you may assume is happening but may not currently fully understand. Note the term "assume" means your hypothesis is not necessarily a factual statement. Your hypothesis should relate to a theory of why you think something you have observed is happening, why it is a problem of interest or why answering the problem would provide your fellow researchers with useful information. The hypothesis could also account for any differences you are trying to establish between two different sets of data.

Review again the professional literature you identified as relating to your problem. Examine the theory or theories the researcher and any assumptions they may have made. Apply this information to finding an explanation (theory) for how your problem can be resolved.

When writing your hypothesis, you start with your problem statement. The wording of the hypothesis should be a statement which brings into focus the ***critical decision or issue*** in your research.

Your statement, hypothesis and the analysis of the data you collect allows you to make the critical decision with a certain level of confidence. Most researchers select a level of confidence of 0.5 (95% confidence that your sample is a valid sample of the population), or the level of confidence may be 0.1 (99% confidence). Note: 99% is a more stringent standard of confidence. If testing the sample data produce results consistent with the research hypothesis, the hypothesis is accepted; if not, it is rejected.

NULL & ALTERNATIVE HYPOTHESES:

We should write two forms of our hypothesis for each issue we are researching. One is written as the null hypothesis where "null" or "nothing" means "no difference." This type of hypothesis is expressed as an equality, but not just by using the term "is equal to." Because some of the statistical tests provide not just a single result but a range of results, the equality may be stated as either: "less than or equal to" (the symbol $\leq$); "greater than or equal to" (the symbol $\geq$); or just "equal to" (the symbol =).The symbols $\leq$ and $\geq$ represent a region under the frequency curve that is bounded by the equality. So, $\leq$-1 refers to the region to the left of and including -1; $\geq$-1 refers to the region to the right of and including -1.

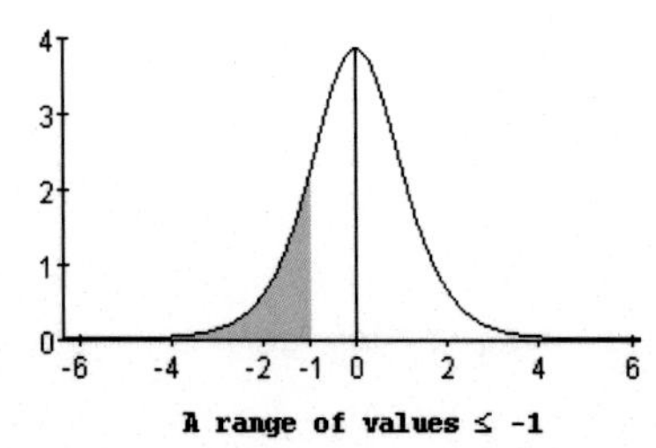

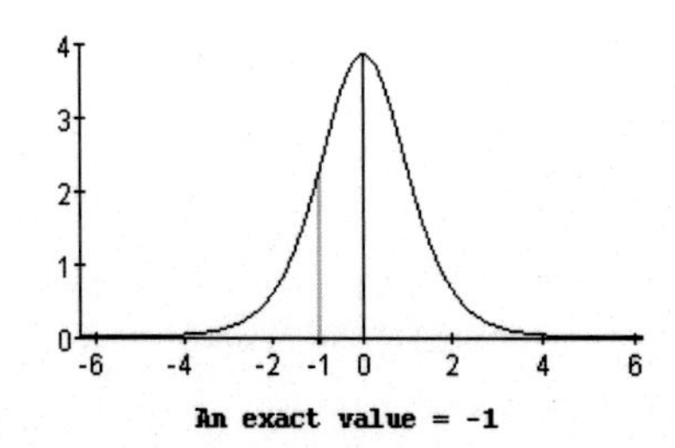

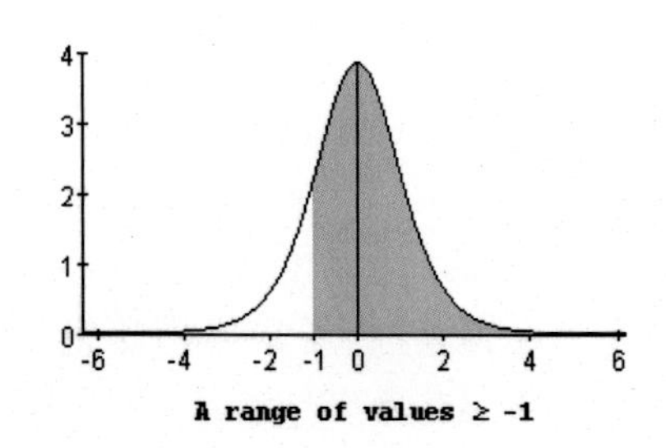

Figure 6: In this example, the red areas under the curve are defined by the null hypothesis with a p-value of -1 (negative one).

The second form of your hypothesis is called the alternative hypothesis. The null hypothesis is written as H_0, and the alternative hypothesis is written as Ha. To explain in more detail:

A formal hypothesis must be written in the form H_0: $P \leq X$, where H_0 is the null hypothesis, P is the parameter of concern and ***X*** is the value of ***P***. The null hypothesis is represented by H_0 (H for hypothesis with a subscript of zero). The equality means "there are no significant differences..." In the figure above the regions shown in red illustrate the ranges formed by the use of the equalities ≤, ≥, or = even though the symbol = represents a single datapoint forming a line under the curve.

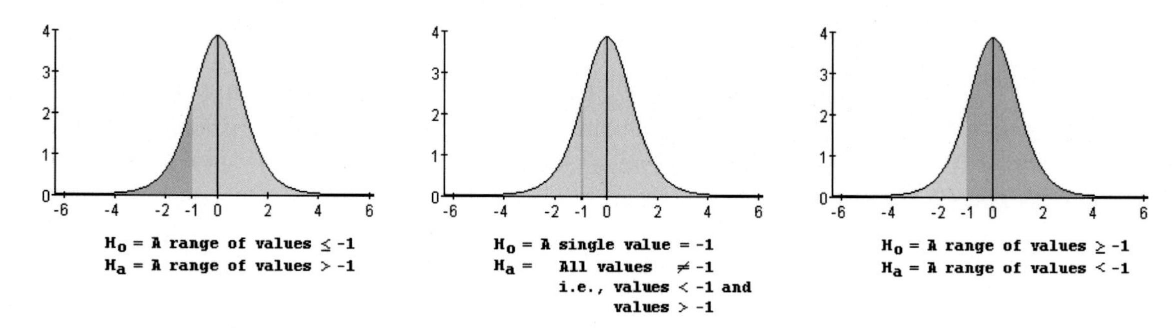

Figure 7: The green areas are defined by the alternative hypothesis with a p-value of -1 (negative one).

AIMS OF RESEARCH:

Generally, the null hypotheses can be stated in the form that either the hypothesis is false or there is no difference. Given The null and the alternative hypotheses, the researcher hopes to prove the null hypothesis is false (that is, it is rejected) or that there is a real difference. When the null hypothesis is rejected, this implies the acceptance of the alternative to the null hypothesis, one which states the opposite, that there ***are*** significant differences. This is the ***alternative hypothesis,*** with the symbol H_a, a statement on which your study is based. In Figure 7 above, the regions shown in green illustrate the ranges formed by the use of > (greater than) or < (less than) or ≠ (not equal to). The green regions are the ***rejection regions*** for the null hypothesis. In other words, if the calculated value lies anywhere in the green regions, the null hypothesis (H_0) is rejected. If it lies in the red regions you cannot reject H_0. ***NOTE:*** H_a is the ***complement*** (makes complete) of H_0, that is, H_a contains all the values under the curve that are not contained in H_0. This means that H_0 and H_a are mutually exclusive, so that if one is true the other is false.

Set	Null hypothesis	Alternative hypothesis	Number of tails
1	$\mu = M$	$\mu \neq M$	2
2	$\mu \geq M$	$\mu < M$	1
3	$\mu \leq M$	$\mu > M$	1

Figure 8: The null hypothesis determines the alternative hypothesis and the number of tails.

The three forms of the null hypothesis are: H_0: $\mu = M$; H_0: $\mu \geq M$; H_0: $\mu \leq M$; and produce inequalities in the alternative hypothesis: H_a: $\mu \neq M$; H_a: $\mu < M$; H_a: $\mu > M$ respectively.

In expressing a hypothesis in null terms, researchers are aiming to have it rejected. For this rejection to be valid there must be convincing proof that the null hypothesis is false. Casual inspection of data may suggest that there are definite differences. However, the researcher must make a test to determine if the observed differences are a result of chance or are "normal" differences within the population being studied or the results are influenced by some non-random cause, resulting in apparently significant differences.

TAILS ON A CURVE:

When working with interval data, you compare averages referred to as "means" using one of several tests. One aim is to "normalize" the dataset so that you can get a normal curve of the distribution. Here is the normal curve and two skewed curves. ***Skewed curves may be skewed to the left or right. The "skewed" end of the curve is referred to as a tail.***

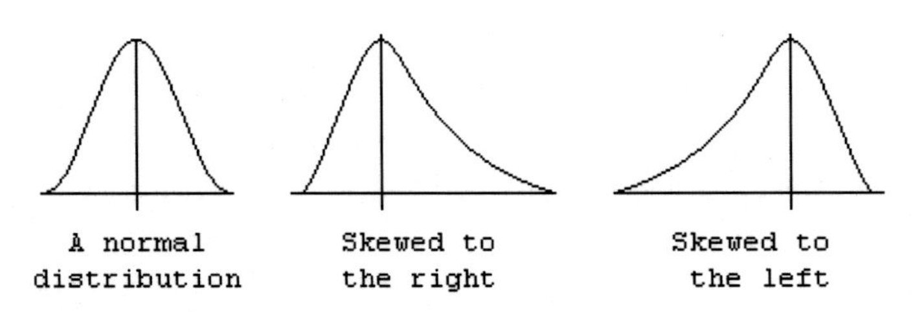

Figure 9: Normal and skewed curves.

One purpose of analyzing data is to examine the tails of the curve. Here are the types of tails that occur depending on the value of ***m (in the example below: positive*** α ***or negative*** α***)***. The number of tails is significant in performing certain statistical tests.

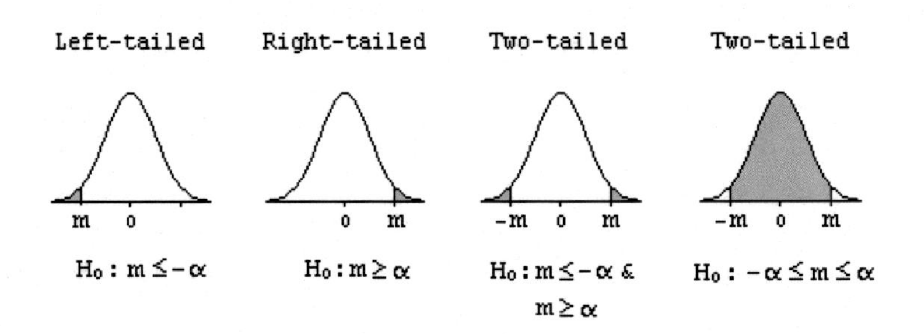

Figure 10: The red areas are the regions of acceptance.

REGION OF ACCEPTANCE:

Suppose we wish to test the population measure M and the mean of the population is μ. Then the null and alternate hypotheses are:

$H_0: \mu = M$

$H_a: \mu \neq M$

If the test statistic lies in the area defined by the null hypothesis, this is the ***region of*** acceptance for the null hypothesis. If the test statistic lies in the area defined by the alternative hypothesis, this is the ***region of rejection*** for the null hypothesis.

TESTS OF HYPOTHESES:

Analyzing the data is part of the research cycle. The hypothesis tells us what classification of data to collect for your study and determines which test to use in analysis. The result of the analysis tells us something about our hypothesis. In fact, the purpose of the statistical tests is to test our hypotheses, to determine if the results of our study show significant differences, in other words, whether chance could account for the apparent differences we may observe.

Generally, there is an observable pattern when we collect data. If the data does not show a pattern, this does not necessarily indicate that there isn't one. We may not see the whole picture unless we examine our data carefully or, to use the proper term, its ***distribution.*** There are a number of tests that can be used on data distributions. But, first things first, you need a hypothesis to direct the collection of data.

TYPE I AND TYPE II ERRORS:

A number of statistical tests have been devised for proving that the null hypothesis is (probably) false. The test or tests we apply show whether the differences are real or whether they are attributable to chance. A casual inspection of our data may suggest that there are definite differences. But before we can state that the hypothesis is true or false, we must make one or more tests to determine if our findings are simply due to chance. Differences we observe may be a result of "normal" differences within the population and there can be errors in analysis.

In performing a test, we determine if the stated hypothesis is true or false. Because statistical tests are not infallible (they are based on probability), two basic errors may occur. These are: A Type I Error and a Type II Error. In general, the larger the number of subjects (***sample size***) in a study, the less likely the researcher is to make either error.

TYPE I ERROR:

We may reject the null hypothesis as ***false*** when in fact it is true. In other words, by rejecting the true hypothesis we make a Type 1 error.

TYPE II ERROR:

We may accept the null hypothesis as ***true,*** when it is not. That is, we fail to reject the null hypothesis when it is false resulting in a Type II error.

HYPOTHESES FOR ONE- & TWO-TAIL TESTS:

When testing interval data, we can make several comparisons of the mean or means of our dataset(s). Suppose the test statistic is ***m***, the value of the test statistic is ***α***.

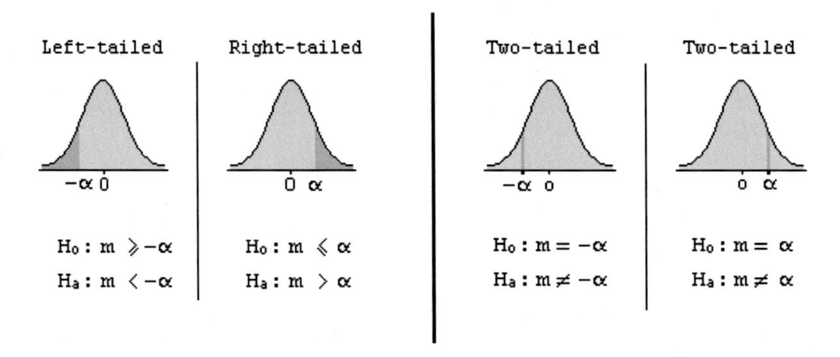

Figure 8: The null and alternative hypotheses for one-tailed and two-tailed tests.

Considering the region of rejection, there are two possibilities: a ***one-tailed test*** or a ***two-tailed test.*** The ***tail*** refers to either end of the distribution curve. ***In the figure above, the null and alternative hypotheses are shown for (left) the one-tail test and (right) for the two-tail test.***

WRITING YOUR HYPOTHESES:

Your problem statement and the information you gained in your literature search help to define the hypotheses you need. Your hypotheses direct the gathering of data. A null hypothesis can be proven ***false*** but a hypothesis written in positive terms ***cannot be proven,*** only shown that it is ***probably true.*** However, we generally act as if our hypothesis is proven. You need to state both the null and alternative hypotheses. These are dictated by your research question and must be stated in such a way that they are mutually exclusive; that way, if one is true the other must be false. Null & Alternative hypotheses are ***not*** two different hypotheses. Think of them as being two sides of a coin. Because H_a is the complement of H_0 (and vice versa) the two together make one whole.

Technically "null" means there is no difference or is equal to something specific; that is the null hypothesis (H_o) always contains an ***equality***, one of: "greater than or equal to" (≥); "less than or equal to" (≥); or simply "equal to." Mutually exclusive means that for: "greater than or equal to," the alternative hypothesis (H_a) must be "less than" (<); for "less than or equal to," H_a must be "greater than" (>); and for "equal to," H_a must be "not equal to" (≠). These three are ***inequalities*** so H_a always contains an inequality. In the following examples, the running time of the small engines in the sample will produce an average or mean (μ). The analysis of the data in each case will produce what is known as a test statistic. The test statistic allows us to determine if the data for the sample involved is or is not significant:

- If the manufacturer claimed that the engine ran continuously for **exactly** 240 minutes, then $\mathbf{H_0}$: $\mu = 240$ and $\mathbf{H_a}$: $\mu \neq 240$.

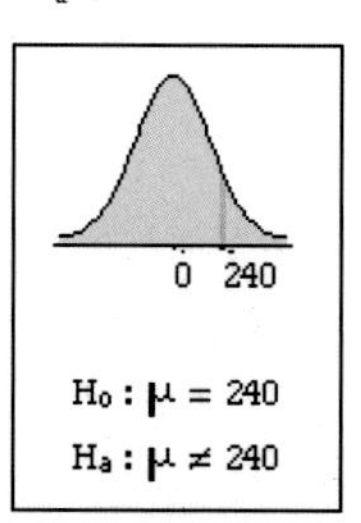

  - With $\mu = 240$, the red line shows the value of μ, and the green areas show the two tails.

- If the manufacturer claimed that the engine ran continuously for **at least** (i.e., more than) 240 minutes, then $\mathbf{H_0}$: $\mu \geq 240$ and $\mathbf{H_a}$: $\mu < 240$.

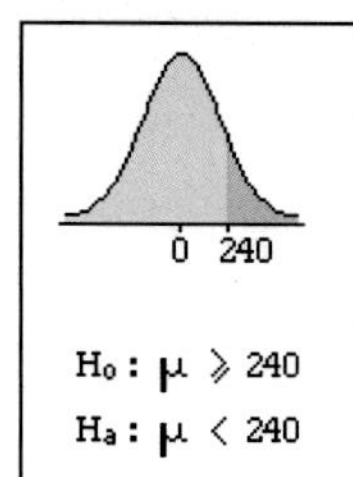

  - With $\mu \geq 240$, the red area shows the range of values of μ. This curve shows a right tail.

- On the other hand, if the manufacturer claimed that the engines ran continuously for **up to** (i.e., not more than) 240 minutes, then $\mathbf{H_0}$: $\mu \leq 240$ and $\mathbf{H_a}$: $\mu > 240$.

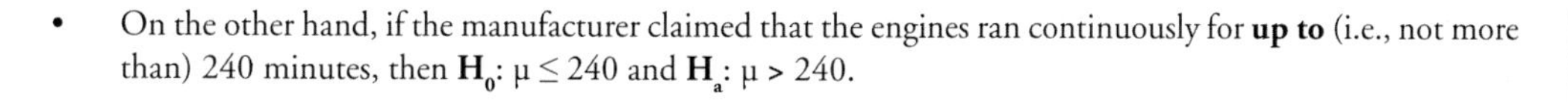

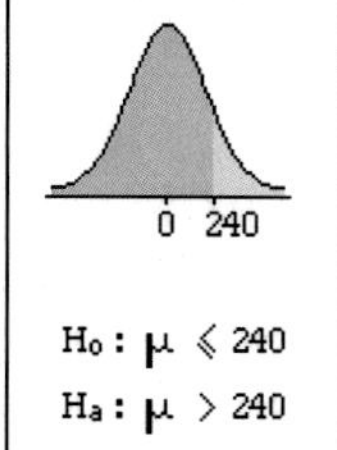

  - With $\mu \leq 240$, the red area shows the range of values of μ. This curve shows a left tail.

The rejection regions are shown in green. These are the regions where H_0 may be rejected. In the first case, we have a two-tailed test and the rejection region is any significant value not equal to the test statistic's value of 240. In the second case we have a one-tailed test, with the rejection region being a tail on the ***left*** as defined by the test statistic's value. In the third case, the rejection region is a tail on the ***right*** as defined by the test statistic's value. *What is also important:* ***all*** *of the values of* H_0 *show equality, whereas* ***none*** *of the* H_a *values show equality.*

UNIT 2B-2 Assignment:

1. **Describe the two types of errors that may be made when testing a hypothesis.**

2. **Take your problem statement and your bibliography and write hypotheses:**

    a. Review your problem statement (think: What do I need to do to solve this problem).

    b. Develop hypotheses, which should define what you believe about your question or problem statement. (Remember the final version of your hypothesis will help you determine your sample size, how you are going to collect the data you need and begin planning the research)

    c. Write your hypotheses as null hypotheses (H0:) and as alternative hypotheses (Ha:).

    d. Analyze these hypotheses and determine what assumptions you may have made, what terms you will need to define, and what data you will need to gather.

    e. Prepare a report for the rest of the members of the class.

3. **Review the reports of the rest of the group, analyze them and write a critique with suggestions for improvement (be prepared to justify your suggestions).**

4. **Participate in a group discussion to critique each student's report; then use their critiques to revise your hypotheses**

5. **Based on the discussion and your real interests in the research select one (1) hypothesis for your study this semester making sure that the data you need is interval data, nominal data, and also ordinal data**

6. **Hand in the critiques given to you, your original, and your final hypothesis.**

UNIT 2C: DATA COLLECTION TECHNIQUES.

PURPOSE:

In this unit, you will complete your preparations and gather the data needed, interval, nominal and ordinal. Note: If your hypothesis does not allow for all three types of data, you will need to change it.

Gathering data is necessary for testing your hypotheses. You are required to do that as part of this unit. The processes for testing that data are discussed in Units 3 & 4. Your hypothesis will allow you to test the differences or similarities between observed and expected results. The e***xpected results*** are those that your theory about your study question/problem statement says you should get. The ***observed results*** are those that come from testing your data. Sometimes they are different!

Now that you know what ***types*** of data you can collect, it is time to learn how to collect it. There are several methods, personal interviews, telephone conversations, sending out a surveyor with some kind of a survey instrument, such as a list of questions, or an actual questionnaire. All these methods seek for answers using a questioning technique. ***Remember, that your hypothesis and the data you need to collect determine how the questions will be phrased. Be observant, the questions are where bias can creep in.***

OBJECTIVES:

- Using your Research Question, Problem Statement and your Hypothesis:
    - Determine what you really need to know about the demographics of your population/sample to achieve your research goals.
    - Determine the information you need to collect in order to achieve the goal of your research.
    - Determine the type of data (nominal, ordinal, and interval) that applies to what you need to know and the information you need to collect it.
    - Determine the best technique(s) for you to use in gathering data for your research study.
- Prepare a survey instrument for use in your research study, include a description of what you hope to achieve with it and how you plan to collect the data.

INTRODUCTION:

When collecting information for a research study, it is probable that data of several kinds will be collected. We may collect nominal by getting demographic information, that is, data by which we can classify the subjects of our study. This data might include age categories, gender, level of education, and so on. We may wish to learn about the respondents' preferences or biases, to better analyze the data with reference to the intent of our study. So we collect ordinal data by asking the subjects to rank certain items, such as work procedures, reading preferences, and so on. We collect interval data, such as the cost of certain items in the budget, or the scores obtained in a test.

NEED FOR PRIVACY:

There is a strong need for privacy in humans. You need to assure them that their names will NOT be used. Explain that the main reason you need their names is so you don't have to say "Hey you!" but otherwise it isn't needed for the study itself. The information they give will be used in statistical analysis but the person responding will remain anonymous.

Any demographic information used will be separated from any identifying feature. THIS IS STRICTLY NECESSARY. Identifying a person in the analysis of data will contaminate the data. If you are obtaining anecdotal information, where there needs to be some means of identifying the individual(s) involved, the normal procedure is to identify them by initials or a false first name. YOU may know who it is, but the people reading your report must not.

IMPLICATIONS FOR GATHERING DATA:

If you set out to gather interval data you will frequently get data that is ordinal and nominal as well. Gathering ordinal data will usually also give nominal information. However, gathering only nominal data gives no other type. The question arises: in future studies, should you gather nominal, ordinal or interval data, two of them or all three? To respond correctly, you must consider the predominant aspect of your study, and the purpose for which the data is to be collected.

DATA TYPE	NATURE OF THE DATA	DESCRIPTION
Nominal	Categories	Classification by attribute
Ordinal	Ranked Order	An ordered sequence
Interval	Numeric values	Values falling in a continuum

Figure 9: The three types of data you are to collect in this study.

If the information you need is quantitative, such as the scores in an exam or a person's performance, and we wish to compare the differences between those values, then you collect interval data. If your major concern is how an item is ranked in comparison with other items in the same category, then the data you want is ordinal. If you wish to compare items that fit in different categories, then you gather nominal data. It is probable that you will gather identifying information (nominal) as well as the interval data. For this study, you are required to collect all three types of data. That still leaves a question: What should you collect?

DESCRIBING THREE TYPES OF RESEARCH DATA:

NOMINAL DATA: You should consider collecting nominal data such as age; gender; physical characteristics, such as height, weight, hair color, etc.; level of education which includes pre-school, kindergarten, grade level, etc.; birth sequence if considering a family; and whatever else applies as long as the data is collected as categories (labels). This data is gathered by "fill in the blanks" at the beginning of any survey instrument.

ORDINAL DATA: This type of data is likely to include preferences of the subjects of your study. Ask yourself: What do I want to know about how the subjects think and feel? Information that could be useful: preferences of programs on T.V.; types of movies or books they like; leisure activities; work preferences; foods; places to eat out; etc. If you have specific items, such as "Aero Chocolate Bars" or places in mind, such as "Joe's Bar & Grill" it is helpful to provide as an additional choice: "don't know" in case any respondent has never heard of it. When you want the subjects to make a choice, the usual method is to present a question about the item/place and provide a set of five preference levels (called a Likert Scale) in a range (1, 2, 3, 4, 5), such as: Like a lot, like somewhat, neutral, don't like, very much dislike. It is also possible to have a ten point scale, but generally the additional choices tend to dilute the results. The choice of "Don't know" is not a preference so while you would count the number who chose that answer, "don't know" would count as zero as far as the preferences are concerned. You can also use a statement about a particular item or place., such as "Final Fantasy X9" is one of the best role-playing games I have ever played." Then the subjects are asked to select: Strongly Agree, Agree, Never Heard of It OR No Preference, Disagree, Strongly disagree or some other five-point scale.

Make sure throughout the questionnaire, that questions of this type are consistent. For all questions with a five-scale set of choices, the rankings should be the same, for example, one (1) as the lowest and five (5) as the highest; ***or*** rankings can go from one (1) as the highest to five (5) as the lowest. One or the other not both! If you mix the rankings so that some have five as the lowest and others have one as the lowest, your data will be unusable.

INTERVAL DATA: Interval data falls in a continuum of numeric values. The data may be discrete, such as the age of freshman college students. You will sort the ages of the students into how many of each age. You may be told that in the past the average age of Freshman students has been 22. You would collect the ages of the incoming freshman and how many fall in each age of, say, 17, 18, 19, and on through 25 or 30, and determine the average age of the current intake to see if the current intake has the same average age as earlier intakes. Such data may be sorted into intervals, for example: less than 17, 17-19, 20-24, 25-29, above 29. The advantage of the intervals is that outliers are included in the "less than" or "more than" so any gross outlier will be included without the value of the outlier contaminating your data spread. When you use intervals, it is recommended to begin each interval with the numbers 5, 10, 15, etc. rather than end them with those numbers.

One common type of data gathered is the scores of exams. We may use individual scores which are likely to be used with tests where the maximum score is 10, or 20. Or we may sort the scores into intervals which are likely to be used for mid-term and finals where the maximum score may be 50 or 100. The students are sorted according to their score using the discrete type of data (for scores ranging to 10 or 20) or in the continuum (for scores totaling 50, or 100).

You could be comparing an individual with him or herself as in a series of tests, or with the rest of the students in the same section or across all sections, comparing how well each individual scores in the midterm or final. Any time you gather numerical data, you are gathering interval data. The commonest methods of gathering data are interviews, surveys or through questionnaires.

INTERVIEW TECHNIQUES FOR GATHERING DATA:

INTERVIEWS:

An interview is used for the purpose of direct questioning of one or more people. For example, you can interview a single person at a time until you have interviewed all subjects; or you can interview the whole group at the same time; or where you have several families as subjects, one household at a time.

You personally need not be the interviewer and when it is a household, you can use an enumerator. It is best to arrange a time for the interview in advance. Whoever, including yourself, does the interviewing needs to have a ***script*** and the interviewer must adhere strictly to that script. The script will probably be in the form of a set of questions. There needs to be space for answers and observations to be written in. At the beginning of the interview, tell the person being interviewed how helpful their answers are expected to be and a little about your goal in doing the interviews. Tell them you are going to take notes of what they say.

A second choice for the interview is to use the telephone or cell phone.When you contact the person, first ask if they have a few minutes to respond. (DON'T call during normal meal-times!) If they don't have time "right now" ask when it would be convenient to call. Explain what you hope to achieve and why their responses will be useful.

INTERVIEWING SUBJECTS, 14 OR 15 AND OLDER:

In the case of a group, you may use, with the group's permission, a tape recorder to record their answers. If you do use a tape recorder, place it where the responses can be heard but where it is less obvious to the subjects. It takes up to four or five minutes for a person to relax if being recorded for the first time. At first, the person doing the interview can talk briefly about your research aims and how helpful their responses will be then have the members of the group respond to general questions about who they are to relax them before beginning on the research questions. Try to keep a relaxed and informal feeling during the interview. A ***focus group*** is a group where the participants are encouraged to talk freely to express ideas or explore attitudes and feeling. This is useful to gather data with the aim of designing your questionnaire for other groups or individuals. If the group gets too far afield from your question, it is alright to bring them back to the subject but keep the attitude light. A sort of "yes, that is interesting, but we need to focus on..."

WHEN USING A SCRIPT: There are occasions when a response to one question suggests another which was not included in the original script. It is recommended to include an empty half-page, or so, at the end of the script for the interviewer to write in any additional questions asked and the responses given to those questions. But it is also recommended that the interviewer write down the question thought of but NOT ask it as that question is outside the scope of the original script.

INTERVIEWING SUBJECTS, 8-14:

Use similar techniques for interviews as they work well with this age group. Although there may be someone who is shy, generally this age group is less inhibited and has fun as long as you can keep the questioning lively. It is important to keep your voice soft and low as the liveliness can get out of control. (One researcher filled in as an instructor of five 13 & 14 year olds for three weeks, later the researcher realized that the only week the group was less rambunctious was the one when she had laryngitis and couldn't speak above a whisper!) It is probably best to not interview this age level in groups as each person in the group tends to egg the others on and that could result in chaos. For this age group, ***follow the script!***

INTERVIEWING SUBJECTS LESS THAN THE AGE OF 8:

It is often helpful to have an assistant when working with children under 8. The extra pair of eyes may be helpful. Subjects in the 4-7 year old age range can be "interviewed" on an individual basis. ***Informal questioning techniques*** to explore their feelings, attitudes, ideas, etc., are acceptable. Laughter is very helpful. Subjects age 3 and under are best observed with the observer being as "invisible" as possible, through a one-way window is best but sitting at a distance from the child, out of his or her view, and after allowing the child to become accustomed to your presence - do not begin the observations while the child is still interested in you or your presence.

Babies and toddlers can be observed and their behavioral responses recorded. Your recording sheet should already have descriptions of what you are going to do, one sheet for each child with room for your observations to be recorded. But you must be ***really alert*** to what is happening. Eyes in the back of your head might help!
Examples of research studies with children in this age group:

1. Observing babies' reactions to photographs of life-sized faces with various expressions, such as smiling, frowning, angry, etc. The assistant would present the photo while you observe the reaction.
2. Observing babies old enough to crawl and placing a variety of toys within reach, sometimes with the bright colors of the toys as a factor in the research, such as, how many children will choose red versus blue. The assistant can draw the child' attention to the objects used in the study.
3. Observing toddlers reactions when placed with another toddler of the same age and some toys are made available.
4. Observing the 4-8 year old child, when you wish to find out something that can be demonstrated during the interview by the assistant, while you observe the child's responses.

These are only some of a number of social actions, reactions and interactions that can be observed with the observer being behind a one-way window. If you have to be in the same room, endeavor to be unobtrusive while recording your data.

GATHERING DATA THROUGH SURVEYS AND QUESTIONNAIRE INSTRUMENTS:

The Survey instrument is a method for obtaining data from individuals through asking specific questions. The questions are set up beforehand. Although the instruments can be used in an interview, generally they are used with other collecting techniques, such as door-to-door, as in previous U.S. Censuses. Much of the questions are to do with demographic data, such as the names of the people in the household and their ages, relationship to the head of the household, marital status, occupations, income which is given as a range, for example, less than $15,000; $15,000-30,000, $30,000-60,000, %60,000-100,000 and so on (notice the unequal intervals); years educated such as 12 years (high school graduate), 16 Years (4 years of college), 20 years (Ph.D. or similar). As with the interview, the surveyor or enumerator must keep to the script.
One important feature of using a survey or questionnaire is the importance of privacy to most people. The respondents need to be assured that they will not be identified in any way. That only the actual answers will be used. Any demographic information used will be separated from any identifying feature.

The questionnaire may be used in mailings where a stamped-addressed envelope is included for the respondents to use in their reply. This method may also use emails. Either has a disadvantage as one has to rely on the people who receive the questionnaire to respond. In addition, the respondent has to keep track of the papers for the questionnaire. And, sometimes, the way those who do not return the questionnaires would have contributed to the study is more useful that those who do!

One group of researchers thought of a way to increase the responses. There was a list of all the people who received the questionnaire with their address and phone number. Each of them was given a code, such as the initials of the subject with two numbers. The return envelopes were marked in invisible ink with the code.

When the return mail was received, the code was used to check off the individuals. After a period of time to allow for laggards (about 2 weeks) the list was checked, and the non-responders were noted and telephoned. They were asked if they'd had time to respond to the questionnaire (as if the researchers didn't know). Then they were told how important their responses were and encouraged to respond. This brought a number of additional responses. Two weeks later those still not responding were called again and again encouraged to respond. No more than the two contacts were used, but the final response was between 75-80%.

When personal contact is not the medium used with the subjects, the chance of non-responders increases. Another approach can be used with the non-responders. Use a door-to-door survey with a new instrument that endeavors to find out why they didn't respond. The researcher hopes to discover if the initial non-responders were different in some way from those who did respond.

More details, articles about surveys, and assistance in creating surveys may be found by pasting the following on the internet search bar:

www.esurveyspro.com

Click on the "Survey Templates" choice on the "Home; Prices; features; Survey Templates, etc., bar." There are several articles on the right side of the first list under the title of "Tips and Tricks." which are helpful.

Under Tips and Tricks is an article titled: *10 Easy Ways to Increase Response Rates.* And a second: *Survey Design: Writing Great Questions for Online Surveys.* This article has some excellent "don'ts" for writing questions for questionnaires (which do not have to be for online surveys only). Click on either title.

TYPES OF SURVEYS:

Surveys usually ask for demographic data in an effort to get characteristics and attributes about a particular population. The data is collected from a sample of the population and the tests of the data allows for an estimate of the population characteristics. There are a number of survey techniques.

Pilot study: using a survey instrument. A pilot study is usually done on a small scale to improve the quality of the survey instrument to be used in a full scale study.
Exploratory survey: Done where little or nothing is known about either the subjects or the domain to be surveyed, and the intent is to get enough information so that a survey instrument may be devised for a full scale study.
Repeated survey: Sometimes you may want to sample survey a population but want to be sure that the sample is a good representative of the population. One way to do that is to survey, using the same instrument, different samples of the whole population.
Analytic survey: Similar to a repeated survey, but there are two or more samples of the population and the intent is to compare the results between the samples.

CONDUCTING A SURVEY:

Suppose you were comparing demographics of two towns in Iowa to determine if their population is similar enough to say they really are "sister" towns. You want to determine if the age group frequencies have the same average. If one town is predominantly composed of young families and the other has many more retirees, your hypothesis would be that the average age for each town would be significantly different. On the other hand, if there is an even mix within each town of all age groups, then the average age for each town would probably be very similar. The null hypothesis would state: The average age of the inhabitants in town A does not significantly differ from the average age of inhabitants in town B. The test or alternate hypothesis would state: There is a significant difference between the average ages of inhabitants in town A and Town B.

The sample size you need will depend on the size of the population. Suppose there are only 500 inhabitants in each town. You'd need to have a sample size of 217 individuals from both A and B. If, instead you were interested in families of two or more people and there were 100 such families in each town, you would need a sample size of 79 families from each town. (Your study would not be adversely affected if one town had, for example, 103 and the other 105 families.) Your method of collecting data would differ if you were looking at individuals or families. For the families your survey instrument would collect names of individuals in each family and their age. You would also have to have some means of identifying which family the data is from.

Note: When taking any kind of survey, it is important to promise anonymity of response. One way this can be done is to mail the questionnaires with a stamped addressed return envelope. Invisible ink can be placed on the return envelope ***(not on any of the questionnaires)*** for identifying which subjects have returned the questionnaires. Once received, the questionnaire is separated from any identifying item so the data is still treated anonymously but you can check with the families who have not responded asking "Have you had time to complete and return your questionnaire?" Then encourage them to do so. It is acceptable to do this no more than twice. Another reason for knowing who has or has not responded is that you may need to find out if the non-responders as a group would have given similar or different responses. The fact they did not respond may be because as a group they were different in some way from the rest of the sample.

TYPES OF QUESTIONS:

In the paragraphs below, some words or phrases have “<” at the beginning and “>” at the end. This represents a phrase you would replace with something appropriate to your study.

Examples of responses are shown below. For your own questions, use whatever you feel is appropriate. Be careful about asking questions that get a yes/no answer. For example, “Do you like <something appropriate>?” could be written so that the responder must rate the attitude. Also open-ended questions are generally not appropriate. If the question is written as: “How do you feel about...” Then the responses could be to make a choice between: “Like it a lot… Like it somewhat… Neither like nor dislike… Dislike it somewhat… Don’t like it at all.” The exact wording of the responses can be varied. But the major question would be more useful if it was worded: “Rate how you feel about ...”

Even if you’re looking for the knowledge of the respondent: “Do you know <such-and-such>?” where <such-and-such> is replaced by something appropriate, allows only three responses: yes/no/don’t know. One might respond: “Yes I do, so what?” It would be better to ask the question so as to get an opinion, or an attitude about what you want to know. However, in some circumstances, there is no other way to ask the question. You may use this question to separate the people being surveyed by a particular attitude or knowledge base. In this case, if there are a series of questions (up to five) on a single topic you may add instructions, for example: “If you chose “Yes” continue with question 10. If you chose “No” skip to question 14. If you chose “don’t know” skip to question 19.

The set of questions 10-13 are different from the set of questions 14-18. Question 19 resumes the general questioning for all respondents. There is no point in having the respondent go through the whole set of questions which are dependent on the response to the first question.

Opinions can be sought by making one or more statements and asking the respondent to: “Definitely Agree, Agree Somewhat, Neither Agree nor Disagree, Disagree Somewhat, Definitely Disagree.”, for each statement. The words may be written under a line which has marks to show the five points of the answers; the marks should be set at equal distances.

Another format for these types of questions would be to ask the respondents to rate a series of questions, explaining that “1 is excellent down to 5 is dreadful.” (Or vice-versa.) Then the numbers 1 2 3 4 5 are written in for each question. Note: keep your response choices consistent, either 1 is *always* the most preferred *or* 5 is; don’t mix them up. Other types of questions can include true/false, multiple-choice and occasionally a brief written answer. But reduce any question requiring a written answer, no matter how brief, to an absolute minimum as people rarely want to write.

THE FORMAT OF THE QUESTIONNAIRE:

A useful article on how to develop and test questionnaires is found at the URL below. It also provides a model for a research article.

http://www.joe.org/joe/2007february/tt2.php

The format of your questionnaire should be consistent across all pages of the document. The question, which can use more than one line, should be in the left portion of the page using between one half to two thirds of the width of the page. The 5-scale responses should be fitted into the other half to one third of the width. NOTE: in the next paragraph are the words: <information being asked for>. The pointy brackets < and > indicate that the items inside the bracket should be replaced by pertinent information based on your research study.

Keep your questions to no more than three double-sided pages. The last part of the last page should allow room for an open-ended response such: "What else would you like to say about the <information being asked for> that wasn't included in this questionnaire?

COMPLETENESS CHECKS:

At the survey level where an enumerator was used, check to make sure that all households have a response even if it is only "No-one was home" or "Refused to respond" or "Couldn't find the place." This information can be used for verification later.

At the survey level, count the number of questionnaires and match to number of respondents to make sure all have been returned even if no data was written into the questionnaire. Individual questions can also be rated if some questions were not responded to. But record which ones were not responded to.

An important statistic, particularly when you have used a mailing for the questionnaires: compare the number of respondents and the number of non-respondents and write as a percentage. This gives you the response rate, hopefully above 50%. When there are some who have not responded, you can use a different technique such as a personal interview or a phone interview. Purpose of this type of follow-up is to determine if the non-responders are different in some way from those who did respond.

COLLECTING DATA:

The purpose in gathering data is to help us determine if the data supports our hypotheses or not. If we don't collect the right kind of data, we will not succeed. What you want to know (hypothesis) will determine what data you will collect.

When examining your hypothesis, you must analyze it to decide what type of data you need. Once you know what you need, you can think about how you will collect it. For example, if your hypothesis is saying that: exam results will not be affected by whether the subject is male of female ***but*** there will be a difference the older and more mature a student is. Among the data you collect, your demographic data questions will ask for name, gender, and age.

Since you will be comparing exam scores, your data must include the scores of the respondents' exams. Are you going to compare scores of the mid-term and the final? Or you may present a version of the final exam at the beginning of the semester and a different version at the end of the semester to determine how much progress has been made.

WAYS OF RECORDING DATA:

1. **INTERVIEW RESPONSES:**
   If you have specific questions on your "script" leave enough room to jot down as much of each response as you can. Recording the responses helps, but you will still have to write the responses down or type them into your computer at some point.
   One way to begin sorting the responses is to "cut & paste" a paper copy, so you can gather all the responses to Q1 in one pile, the same for Q2, Q3, etc. But before separating any one individual's answers, write a code beside each question so even when the questions are in separate piles, you can still identify all the answers that go with "Mary" or "John" or whoever. This also applies to any data that is recorded for questionnaires, surveys, etc.

2. **OBSERVATIONS:**
   As with the questionnaires, you should record each individual's response on a separate sheet, coding the different questions so you will be able to identify each individual's responses as well as the total of question responses.

CAUTION 1: THERE ARE NO FOOLPROOF QUESTIONNAIRES!!!

No matter how carefully you work on creating the questions for your questionnaire, or how sure you are that none of the questions can be misinterpreted, it is almost inevitable that you will find at least one respondent who will indeed misinterpret at least one question in your survey.

Record the information when this happens!

CAUTION 2: BEWARE OF BIAS.

Just because your theory is that a "certain thing is so," does not mean that everybody feels the same way. Be careful to design your questionnaire so that it does NOT become clear what answers you are looking for! For your survey to be valid there must be no bias, visible or invisible! One way to eliminate bias is to take your questions to a small group of, possibly, fellow students and ask them to critique your questionnaire.

UNIT 2C Assignment:

1. **Before you begin your assignment, review all the articles found on your literature search with the following questions in mind about the use of Survey and Questionnaire instruments.**

    a. How did the authors collect their data? If their method is different from what is discussed in this unit, describe it.

    b. Did the authors have any problems with gathering the data?

2. **Then complete the following:**

    a. Use the URL: www/esurveyspro.com/ and review the section titled "*Survey Design: Writing Great Questions for Online Surveys.*" Write a summary of the information presented in this article.

    b. Using your Research Question, Problem Statement and your Hypothesis determine what you really need to know about your population/sample and the type of data (nominal, ordinal, and interval) you must collect to achieve the goals of your research.

    c. Prepare a survey instrument for use in your research study, include a description of what you hope to achieve with it and what techniques you plan to use to collect the data.

    d. Provide copies for your discussion group and collect those of the other students, then write critiques. Be on the lookout for information in the survey that shows bias or that could be misinterpreted!

    e. Discuss the questionnaires and critiques in the discussion group.

    f. Hand in your original, the critiques and the improved instrument.

UNIT 2D: WRITING A RESEARCH PROPOSAL.

PURPOSE:

Whenever someone plans to do research, the funding for the project must be found. This takes the form of a proposal which shows what is to be researched and its importance, the hypotheses directing the research, what the researcher expects to discover, and an indication of how much it is going to cost (time, materials, transportation needs, etc.). To complete the requirement of this unit, you must make a proposal to your department chair and your faculty advisor. It will be submitted once you have had opportunity to improve it with the help of your discussion group.

Before you can begin the study part of your research, there are a number of steps to complete and planning carefully is essential. Some of the planning you have already done prior to this point. The next stage is to use all you have done so far to assist you in writing your proposal.

OBJECTIVES:

- Using all the information you have learned so far in SECTIONS 1 and 2 write up a proposal for doing your research including the justification of your hypothesis.
    - Describe the data you plan to collect and make sure your plan requires you to collect all three types of data (interval, nominal and ordinal).
    - Participate in a group discussion.
    - Provide a copy of your proposal for each student in the discussion group a week in advance.
- Write a short critique of each of the proposals from other students and be prepared to discuss your critiques including why the points you make are valid.
- Use the critiques and the discussion to finalize your hypotheses and your proposal.
- Submit your proposal to your Instructor, your Department Chair and your faculty advisor.

INTRODUCTION:

Review your hypothesis to determine the ***population*** for your study, the ***sample size,*** and the ***type of data*** you will collect. You must be certain that the data you collect will enable you to reach valid conclusions. This takes planning. It is essential to know what planning must be done before you begin. The plan for your research will be achieved by writing a proposal to your department chairman and your faculty advisor asking for funding for your research. [Note: this is a practice activity, so you probably will not need the funding but the proposal must be submitted. You should discuss this with your Chair and faculty advisor beforehand and request them to review it for you!]

ORGANIZING THE RESEARCH PROPOSAL:

When you start to work on your thesis or dissertation, you will need to work closely with your faculty advisor. Before you can plunge into the study on which your thesis or dissertation will be based, you will need to provide your faculty advisor with a proposal. The way the proposal differs from your final research report is that the proposal tells what you plan to do and the report builds on that by telling what you have done. You will easily adapt your proposal, add the description of the research process used, describe the actual data collection, data analysis and discuss how you arrived at your conclusions. The first units in this course have helped you with a number of "pieces" that will be used in writing your proposal. Applicable objectives describe those pieces:

- **UNIT 1A:** Properly define a research problem applying to your major and write it as a Problem Statement.
- **UNIT 1A:** Justify your research into this problem explaining what and why.
- **UNIT 1A, 1B, & 1D:** Describe the methodology you plan to use and write it up as an Introduction to your research report (the proposal should include your problem statement and the justification for doing the research).
- **UNIT 1B:** Determine population and sample size you need to complete your study.
- **UNIT 1C:** Determine what data you will need to collect, identifying it as nominal, ordinal, and/or interval data, answering "What do you want to know and how can you find out?"
- **UNIT 1D & 1E:** Describe how you plan to do the research.
- **UNIT 2A:** Complete a search of the Professional Literature using your problem statement as a guide evaluate the articles you find in terms of their thesis, methodology, and results.
- **UNIT 2A:** Prepare an annotated bibliography of all papers found and select those articles that apply directly to your problem statement. Then make a simple bibliography of your selected articles using your department's style.
- **UNIT 1A & 2A & 2B:** Use your problem statement to write your research study's hypothesis.

CONTENTS OF THE RESEARCH PROPOSAL:

Matthew McGranaghan at the University of Hawaii has put on the Internet "Guidelines on writing a research proposal." the URL to get to this document is

http://www2.hawaii.edu/~matt/proposal.html

Read the whole article, it doesn't have that many pages. McGranaghan refers to two models for writing a research proposal, one is a two-page model and the other is the standard (longer) model, which can be from five to fifteen pages. The two-page model (about five paragraphs) is what you should model your research proposal on. The suggested content of the proposal is summarized here:

Introduction:

Identify the general topic area, present the research question, and establish its significance.

One to Two Paragraphs:

Briefly review the professional literature relating to the topic, identifying who has written on your topic and what they have found (allow one sentences per important person or finding or open questions that come out of their research). Restate your question in the context of these findings and show how it (your question) fits into the big picture.

Methodology:

Describe your methodology: how you are going to approach your research, describe the data you plan to gather, and what you will need in order to do it. This is where you would outline any expenses needed for the research.

Results:

Here you briefly describe your expected results, how you expect to interpret them and how they will increase your colleague's understanding of the "big picture", that is, your addition to the professional literature.

YOU HAVE EVERYTHING YOU NEED TO WRITE THE PROPOSAL!

Go for it!

COLLECTING *YOUR* DATA FOR YOUR STUDY:

While waiting for approval from your Department Chair and your faculty advisor, start making arrangements for your study particularly for collecting your data. ***It will take time to make all the arrangements so the sooner you start the better.*** Work with the information in UNIT 2C: Collecting Your Data, to prepare your methods such as questionnaires for collecting data and begin the actual collection.

Your plan should include no more than two weeks to complete the collection of your data. If you use a questionnaire, keep it short (justify using more than ten questions). Plan a method for collecting responses that does not require you to use the mail.

UNIT 2D Assignment:

1. **Review the article written by Matthew McGranaghan at the URL:**

   http://www2.hawaii.edu/~matt/proposal.html.

2. **Using his guidelines and the summary of his two-page, five-paragraph model on page 3, write a proposal of no more than four to five double-spaced paragraphs. The report you made in Unit 1C may be adapted for your proposal *but assume that you are submitting the proposal to your department Chair to obtain funding for your research.* The proposal should contain all of the following:**

   a. Make a clear statement of your question/problem statement and what you aim to achieve by studying that question.

   b. Describe and justify the use of your chosen methodology.

   c. Describe your hypothesis and justify its value in expanding your major field's knowledge base.

   d. Include at the end of your proposal a bibliography of the articles you extracted as those that will be useful for your study (make sure you use the style required by your department).

   e. Describe what you learned in your literature search referring to your bibliography and include how the search helped define your problem statement and/or question or helped to focus the aims of you planned research.

   f. Describe how you plan to collect and analyze the data and what you will need (survey instrument or questionnaire, or other material) to complete this portion of your study. Outline any needed expenses. If your questionnaire is complete add it as an appendix to your proposal.

   g. Review the proposals of your discussion group, analyze them and determine if they convinced you that funding ought to be provided and write a critique with suggestions for improvement (be prepared to justify your suggestions).

   h. Participate in a group discussion to critique each student's proposal, then use their critiques to revise your own.

   i. Hand in the critiques given to you, your original, and the final version of your proposal.

3. **Present a copy to your Department Chair and your faculty advisor and request their critical appraisal.**

SECTION 3: PREPARING FOR DATA ANALYSIS.

PURPOSE:

Research is the science of collecting, classifying, tabulating and testing data so that significant information can be presented about a given subject. At this point, you should have begun collecting your data. In this unit, you will finish collecting data, complete your preparations and begin the process of testing your hypothesis.

There will be plenty of practice activities for each test you learn and NONE of the calculations will be complicated. Algorithms are included giving a step-by-step process for each test. Just follow the steps carefully and you should have no problems. ***Do not start testing your data until you have collected all of it.***

Since research epitomizes the science of collecting, classifying, tabulating and testing data it is important that you plan the data analysis well. You need to be aware that there are certain types of errors which may creep; there are decision rules that help avoid those errors; and there is a four-step plan to enable you to make sure you not only avoid these errors, but also to ensure that you properly analyze your data and correctly interpret the results.

OBJECTIVES:

- Describe the TYPE I and TYPE II errors and how to avoid them.
- Describe the basics of Probability as it applies to analysis of data.
- Discuss how frequency distributions help in the analysis of data.
- Describe and use the four steps to analyze your data.
- Discuss the measures of central tendency and how to calculate them.
- Discuss the measures of spread and how to calculate the Variance and the Standard Deviation.

PRELIMINARIES FOR ANALYSIZING DATA:

Preparing to Analyze Your Data covers what one might consider "background" information as far as analysis of hypothesis is concerned; the groundwork for the various methods for testing your hypotheses. Topics include "Decision Errors," "Probability," "Decision Rules," "Four Steps for Testing Hypotheses," "Measures of Central Tendency," and "Measures of Spread." These topics must be understood before you can begin actually testing your data.

You need to develop a plan for the analysis of data you collect. There are two aspects of this plan that need to be considered. The first aspect concerns the two decision errors that can affect your analysis. These are referred to as Type I and Type II errors. The first error: rejecting the hypothesis as false when in fact it is true. The second error: accepting the hypothesis as true when in fact it is false. One way to avoid these errors is to draw a curve of your data and determine the regions of acceptance and of rejection. There are also decision rules using the P-value and the Region of Acceptance. These topics are discussed in Unit 3A-1.

The second aspect concerns your plan for analyzing your data. Your plan should include four steps. These are: 1. State the hypothesis; 2. Formulate an analysis plan; 3. Analyze sample data; 4. Interpret the results. The steps are discussed in detail in Unit 3A-2.

UNIT 3A-1: DECISION ERRORS: Type I & Type II.

PURPOSE:

There are two types of error that may occur when testing a hypothesis. See the article on Wikipedia about the Type I and Type II statistical errors, found at the following URL:

http://en.wikipedia.org/wiki/Type_I_and_type_II_errors

The graphics used in this unit were based on some in an earlier article in Wikipedia then adapted to better illustrate the difference between Type I and Type II errors. Type I errors occur when the null hypothesis ***is rejected*** i.e., is accepted as ***false*** when it is actually true. By definition this is a ***false positive.*** Type II errors occur when the null hypothesis ***fails to be rejected*** i.e., is accepted as ***true*** when it is actually false. By definition this is a ***false negative.***

Your analysis plan (step 2 of the general procedure for testing hypotheses described in Unit 3A-1) needs to include the decision rules for accepting or rejecting the null hypothesis. These decision rules are related to the making of Type I or Type II errors.

OBJECTIVES:

- Describe the two types of errors that may be made when analyzing data.
- Describe the decision rules that enable the researcher to accept or reject the null hypothesis.

INTRODUCTION:

Analysis of hypotheses is based on probability. While we can never be 100 % certain that our hypothesis is true, we can be either 90% or 95% or even as much as 99% certain. A number of statistical tests have been devised for proving that the null hypothesis is, probably, false. We want to be sure that the test or tests we apply show whether the differences are real or whether they are attributable to chance. A casual inspection of our data may suggest that there are definite differences; however, appearances may be deceiving. If the data is poorly analyzed, errors may creep in. Before we can state that the hypothesis is true or false, we must make tests to determine if our findings are simply ***due to chance***, or the ***differences we observe*** may be a result of "normal" differences within the population, or ***there can be errors in our calculations.***

TYPE I AND TYPE II ERRORS.

In performing a statistical test, we determine if the stated hypothesis is true or false. Because the various statistical tests are not infallible since they are based on probability, two basic errors may occur. These are: a Type I Error and a Type II Error. Generally, the larger the sample size you have in your study, the less the likelihood of making either error.

TYPE I ERROR:

We may reject the null hypothesis as false when in fact it is true. In other words, by rejecting the true hypothesis we make a Type I error.

TYPE II ERROR:

We may accept the null hypothesis as ***true,*** when it is not. That is, we fail to reject the null hypothesis when it is false resulting in a Type II error.

Any hypothesis contains the possibility of making either a type 1 or a type 2 error. To illustrate this, the next pages show two possible tests: "Not Guilty" or "Not Innocent.

TESTING THE HYPOTHESIS OF "NOT GUILTY":

In a test for "Guilty" or "Not Guilty," H_0 = Not Guilty. Using this test, if a person is judged guilty, then we reject the null hypothesis as false. We are saying the alternative hypothesis, H_a = Guilty, is true. The actual status of the accused and the test results are shown here.

<table>
<tr><td colspan="2">TEST: GUILTY or NOT GUILTY
H_0 = Not Guilty
H_a = Guilty</td><td colspan="2">ACTUAL STATUS</td></tr>
<tr><td colspan="2"></td><td>GUILTY OF CRIME</td><td>NOT GUILTY</td></tr>
<tr><td rowspan="2">TEST RESULTS</td><td>JUDGED GUILTY
(H_0 is rejected)</td><td>True Positive</td><td>False Positive
(Convicted But Not Guilty)
Type I Error</td></tr>
<tr><td>JUDGED NOT GUILTY
(H_0 is accepted)</td><td>False Negative
(Guilt Not Detected)
Type II Error</td><td>True Negative</td></tr>
</table>

Figure 10: Accepting or rejecting the "Not Guilty" hypothesis.

When judged "guilty" the null hypothesis is rejected, that is, we are saying that the null hypothesis is false and the alternative hypothesis of "guilty" is true.

1. If the person is, in fact, guilty of the crime, then by **rejecting the null hypothesis,** we get a **true positive** result.
2. If, however, the person is **not** guilty, then the **hypothesis is rejected when it is true**, giving a **false positive** and a **Type I error.**

On the other hand, if the person is judged "Not Guilty" then the null hypothesis is accepted as true, that is, we ***fail to reject*** the null hypothesis. We are saying that the alternative hypothesis, H_a = Guilty, is false.

1. If the person is **not** guilty, then the null hypothesis is **true.** By not rejecting it, we get a **true negative.**
2. However, if the person **is** guilty, the null hypothesis is **false.** Thus **failing to reject the null hypothesis** gives us a **false negative** or **Type II error.**

TESTING THE HYPOTHESIS OF "NOT INNOCENT":

The other possibility in the analysis is that we test for innocence. This figure shows that, in the test for Innocent/Not Innocent, the null hypothesis, H_0 is Not Innocent. The actual status of the accused and the tests results are shown.

TEST: INNOCENT OR NOT INNOCENT H_0 = Not Innocent H_a = Innocent		ACTUAL STATUS	
		INNOCENT	NOT INNOCENT
TEST RESULTS	JUDGED INNOCENT (H_0 is rejected)	True Positive	False Positive (Guilt Not Detected) Type I Error
	JUDGED NOT INNOCENT (H_0 is accepted)	False Negative (Convicted But Innocent) Type II Error	True Negative

Figure 114: Accept or reject the "Not Innocent" hypothesis.

If a person is judged "Innocent," then the null hypothesis is ***rejected***, that is, we are saying the hypothesis is false and the alternative hypothesis, H_a = Innocent, is true.

1. If the person is indeed innocent, the result is a **true positive.**

2. If the person is **not** innocent, then we have a **false positive** as we rejected the null hypothesis (not innocent) **when it was true.** This is a **Type I error**.

If the person is judged "Not Innocent," the null hypothesis is accepted as true and the alternative H_a = Not Innocent, is regarded as false.

1. If the person is indeed not innocent, we have a **true negative**.

2. If the person **is** innocent, then we have **failed to reject** the null hypothesis **when it is false**. So we have a **false negative** and a **Type II error**.

Be prepared for the possibility of making one or other error in your research.

DECISION ERRORS:

Because we are deciding whether or not to reject a null hypothesis, these errors are referred to as ***decision errors.***

- Rejecting the null hypothesis when it is true gives a false positive which is a Type I error.
- F*ailing to reject the null hypothesis when it is false gives a false negative which is a Type II error.*

THE PROBABILITY OF MAKING TYPE I OR TYPE II ERRORS:

The probability of making a Type I error is called ***alpha*** (often shown using the Greek lower case letter **α**). This probability, α**,** is referred to as the ***critical p-value***. The p-value is the probability that the observed event would occur if the null hypothesis were true. The probability is tested at a significance level of either 0.05 or 0.01. The probability $P = 0.05$ means that there is a 95% chance of not making the Type 1 error and $P = 0.01$ means that there is a 99% chance of not making the Type I error.

Note that in statistics, the term "significance" means that: ***the hypothesis is probably true rather than the result of chance.*** With a significance level of, say, 0.05 or 0.01, the calculated P-value is used to determine if that value is no more than the significance level. The smaller the P-value, the more significant the result is likely to be. If the p-value is smaller than the significance level, the smaller the chance of making the error; that is, the probability of making the Type I error is ***no more than*** the selected probability. For most research studies, $P = 0.05$ is considered an adequate test for ***significance.***

The probability of making a Type II error is called ***beta*** often shown using the Greek lower case letter β. The probability, β**,** of ***not*** making this error is referred to as the ***power*** of the test. The power of the test is the probability that the test will reject a false null hypothesis, that is, the test will not make a Type II error. As the power increases, the probability of a Type II error decreases. The power is defined as $1 - \beta$, so the smaller β is, the greater the power of the test; β gets smaller as ***n***, the number of observations, increases. In other words, the larger your sample, the less likely your test statistic will be subject to a Type II error.

FACTORS AFFECTING POWER[3]:

The greater the power of your test the less likely you are to make a Type II error. In using the power to avoid the Type II Error, you determine the ***effect size*** which is the difference between the true value and the hypothesized value in your study. Suppose you are working with I.Q. data, where the true average value found over a period of many years, is 100. Suppose the sample you have is assumed to have below average I.Q. So your hypothesis might be $H_0 = 90$, and the effect size is 100 - 90 or 10. On the other hand, suppose the sample you have is assumed to be above average, your hypothesis might be $H_0 = 110$. The effect size would be 100-110 = -10.

3 http://stattrek.com/Lesson5/Power.aspx?Tutorial=Stathttp://stattrek.com/Lesson5/Power.aspx?Tutorial=Stat The source for the information on the power of a test is "Tutorial: Lesson 5. Power" Permission to use this information has been given by the author, Harvey Berman.

There are three factors that affect power:

1. **SAMPLE SIZE (N).** Generally the greater the sample size, the greater the power of the test.

2. **SIGNIFICANCE LEVEL (A).** Again in general, the greater the significance level the greater the power level. A value of 0.01 for the significance level is greater than a value of 0.05 which in turn is greater than a significance level of 0.1. Also, if you **decrease** the significance level (for example, to 0.01) you also **reduce** the region of acceptance. As a result you also increase the chance of rejecting the null hypothesis. Therefore, you are less likely to accept the null hypothesis when it is false.

3. **THE "TRUE" VALUE OF THE HYPOTHESIS BEING TESTED.** Note that: the greater the difference between the true value (the parameter of the population) and the hypothesized value (the value for your sample), the greater the power of the test.

DECISION RULES:

In practice, statisticians describe these decision rules in two ways: 1) with reference to a P-value or 2) with reference to a region of acceptance. The source of the following rules is Lesson 5, Hypothesis Testing, used with the permission of author Harvey Berman. The URL is:

http://stattrek.com/Lesson5/HypothesisTesting.aspx?Tutorial=Stat

- **P-VALUE:**

  The strength of evidence in support of a null hypothesis is measured by the P-value. Suppose the test statistic is equal to **S**. The P-value is the probability of observing a test statistic as extreme as S, assuming the null hypothesis is true. If the P-value is less than the significance level, we reject the null hypothesis.

- **REGION OF ACCEPTANCE:**

  The **region of acceptance** is a range of values and a normal distribution curve is used to illustrate the region. If the test statistic falls within the region of acceptance, the null hypothesis is accepted. The region of acceptance is defined so that the chance of making a Type I error is equal to the significance level.
  The set of values **outside** the region of acceptance is called the **region of rejection**. If the test statistic falls within the region of rejection, the null hypothesis is rejected. In such cases, we say that the hypothesis has been rejected at the level of significance.

USING A P-VALUE:

The most commonly used data on probability comes from tossing dice or coins. Coins are simpler as there are only two sides to consider, heads or tails. Statistically, one assumes that there is an equal chance of either one occurring. This also assumes that the coin being tossed is perfectly balanced.

With these assumptions, the null hypothesis would be that 50% of the time, a head would result from the toss, that is, ***H_0: P = 0.5.*** We deal with ***P*** (upper case, representing a population parameter) and write it as 0.5 rather than 50%. The alternative hypothesis is that we get a random mix of head and tails, that is, heads do not turn up 50% of the time, or ***H_a: P ≠ 0.5***; in this case, heads is ***not equal*** to 50%, where ***P*** is the ***hypothesized value of population proportion*** in the null hypothesis. In testing the hypothesis, the ***p-value, p*** (lower case)**,** is the ***sample proportion***, with ***n*** representing the sample size. In this example, the researcher has pre-determined the ***significance level*** for the test, ***0.1***.

Although ***P*** represents a population parameter, we cannot test the whole population of coin tosses! So, to test these hypotheses, we would toss the coin a number of times to get a sample, the sample parameter is ***p***. Suppose we flipped the coin 50 times, resulting in 42 Heads and 8 Tails. Given this result, we would be inclined to reject the null hypothesis and accept the alternative hypothesis. Our conclusion would be that the coin we used was not balanced correctly or was heavier on the heads side, resulting in more heads than tails.

It is important to note the ***probability*** of a head turning up is ***the same each time you toss the coin***! What this indicates is that if you toss the coin and get a head, the probability of getting a head the next time you toss is still 50%! Now, the more times the coin is tossed, the more likely you are to get an even distribution of heads to tails. So if you tossed the coin 500 or 1,000 times, the distribution (proportion) of heads to tails would probably be much closer to 50:50.

REGIONS OF ACCEPTANCE OR REJECTION:

Suppose we wish to test the population measure α and the mean of the population is μ. Then the null and alternate hypotheses are:

H_0: μ = Region of acceptance of null hypothesis.

H_a: μ ≠ Region of rejection of null hypothesis.

Set	Null hypothesis	Alternative hypothesis	Number of tails
1	$\mu = \alpha$	$\mu \neq \alpha$	2
2	$\mu \geq \alpha$	$\mu < \alpha$	1
3	$\mu \leq \alpha$	$\mu > \alpha$	1

The symbol ≠ means "not equal to." The "not equal" symbol represents only one case for the region of acceptance. Here are the three possibilities (Sets 1, 2, and 3).

Figure 12: The null hypothesis determines the alternative hypothesis and number of tails in the curve.

Note that for Set 1, the alternative hypothesis is $\mu \neq \alpha$. This can be interpreted as: any value other than alpha and signifies that the null hypothesis can be rejected. For the other two sets, the rejection region is more complicated. All rejection regions can best be shown with an illustration:

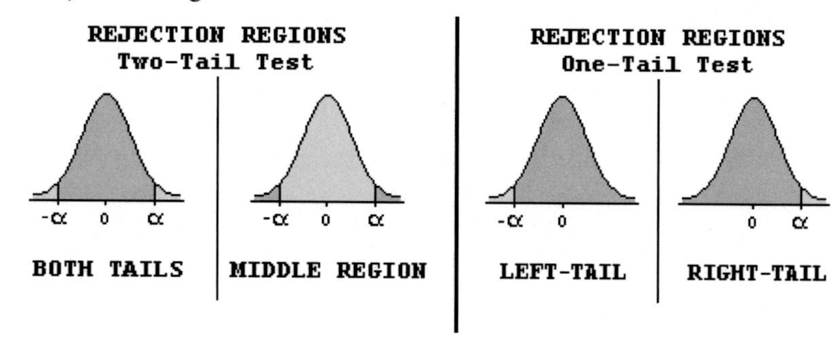

Figure 13: Regions of rejection for two- and one-tail tests.

When developing a research study, you must be certain that your hypothesis is clear and pertinent to your question or problem statement because that hypothesis will direct your research study. Your hypothesis will also help determine the ***population*** for your study, the ***sample size,*** and the ***type of data*** you collect. In a similar way, population, sample size and type of data help to define your hypothesis. You must be certain that the data you collect will enable you to reach valid conclusions. Note: there is a possibility that any apparent differences you find are not significant but the result of chance. You need to be aware of the possible effects of chance and probability.

The curve that can be drawn to represent the data requires that the values of α be positive or negative. The rejection regions are green. In actual fact, Set 1's null hypothesis can be expanded: H_0: $-\alpha < \mu < +\alpha$, the left curve of the two tail test. This is interpreted as "the values of μ lie between –α & +α" (the red region). The right curve of the two tail test show H_0: $-\alpha > \mu$ and $\mu > +\alpha$. The values of μ in this case lie outside the values of –α & +α, the red regions in the curve.

Considering the regions of acceptance/rejection, there are two possibilities: a ***one-tailed test*** or a ***two-tailed test.*** To test your hypothesis, you calculate a test statistic and use it to determine either the ***p-value*** or the ***region of acceptance***. The ***p-value*** is used to determine the strength of the evidence in support of the null hypothesis. If the p-value is equal to the significance level, in this example, 0.1, then the null hypothesis is accepted. If it is ***not*** equal to the significance level, then the null hypothesis is rejected. The process for determining the p-value is shown when the calculations for the various test statistics are discussed (Unit 4: *Procedures for Testing Hypotheses*).

The region of acceptance is determined by the test statistic and is associated with tails of the curve. As mentioned earlier, there may be a single tail, either on the left or the right side of the curve; or there may be two tails one on either side of the curve, as shown in Figure 15.

UNIT 3A-1 Assignment:

1. **Describe the two types of errors that may be made when analyzing data.**

2. **Describe the decision rules that enable the researcher to accept or reject the null hypothesis.**

UNIT 3A-1 Assignment Feedback:

1. **Describe the two types of errors that may be made when analyzing data.**

    - Rejecting the null hypothesis when it is true gives a false positive or Type I error.
    - Failing to reject the null hypothesis when it is false gives a false negative or Type II error.

2. **Describe the decision rules that enable the researcher to accept or reject the null hypothesis.**

    - ***P-VALUE:***

      The P-value measures the strength of evidence in support of a null hypothesis. The P-value is the probability of observing a test statistic as extreme as S, assuming the null hypothesis is true. If the P-value is less than the significance level, we reject the null hypothesis.

    - ***REGION OF ACCEPTANCE:***

      The **region of acceptance** is a range of values defined by a normal distribution curve. If the test statistic falls within the region of acceptance, the null hypothesis is accepted. The region of acceptance is defined so that the chance of making a Type I error is equal to the significance level.

      The set of values **outside** the region of acceptance is called the **region of rejection**. If the test statistic falls within the region of rejection, the null hypothesis is rejected. In such cases, we say that the hypothesis has been rejected at the level of significance.

UNTI 3A-2: FOUR STEPS FOR TESTING HYPOTHESES.

PURPOSE:

During every Research Project, it is vital to test each hypothesis to determine if there is or is not significance. Hypotheses tests differ depending on which type of data you collect. No matter what test you will use, there is a pattern you should follow which applies to ***all*** hypothesis testing. Of particular importance are the decision rules you must make and the types of errors that may be inherent in the testing process.

OBJECTIVES:

- Describe each of the four steps for testing hypotheses and explain their value for any research project.
- Define the four basic tests for analyzing data.
- Describe how the four steps in the process of testing hypotheses will be of value in **your** research project.
- Describe the decision rules that enable **you** to accept or reject the null hypothesis.
- Describe the major types of error that may be made
- Given the name of a particular analytical test (see flowchart next page), identify:
    - The type of data (nominal, ordinal, or interval) to be tested.
    - The type of test (mean, proportion, or relationship).
    - The number of variables involved (one or two).
    - The name of the test (z-test, t-test, Pearson's Correlations test, Spearman's Correlation test, the Chi-Square test).
- As soon as you have collected all your data, take your hypothesis and complete the four steps for testing your hypothesis as you work through the various units describing the tests.

INTRODUCTION:

The stattrek.com[4] site recommends a 4-step process for testing hypotheses. The process:

1. State the hypothesis;
2. Formulate an analysis plan;
3. Analyze sample data;
4. Interpret the results.

This process is useful in that it ensures you don't miss one or more important steps in getting a handle on your data. When conducting research, the four steps should be followed each time you complete a research project. The main purpose of your research is to reach valid conclusions that you can share with other researchers in your field. Using this process will make sure your conclusions are valid and more helpful to your colleagues.

PROCESS FOR TESTING HYPOTHESES:

Let's expand on the outline of the four steps given above.

STEP 1: STATE THE HYPOTHESES.

You need to state both the null ***and*** alternative hypotheses. These are dictated by your research question and must be stated in such a way that they are mutually exclusive; that way, if one is true the other must be false. This step is essential since it is the null hypothesis that you need to test your data but having the alternative hypothesis clarifies what you are hoping to achieve. Technically "null" means there is no difference. In words, we use "the same." So, ***H_0 always includes some form of an equality*** that is each null hypothesis contains the words "is equal to." If the hypothesis is in mathematical form, the symbols **"≥"** for "less than or equal to"; **"≤"** for "greater than or equal to"; or **"="** for "equal to." On the other hand, the alternate hypothesis, ***H_a***, ***never*** has "is equal to." If the alternate hypothesis should be worded so that it does have "is equal to" it is NOT an alternate hypothesis and must be edited. The written form of ***H_a always includes an inequality,*** < for "less than"; > ***for*** "greater than" and ≠ ***for*** "not equal to."

The wording of the hypothesis is important. Examples:

- An inventor of a filling machine for gallon-sized ice cream containers claims their newest machine can hold enough ice cream to fill **exactly** 500 gallons without needing a refill. The null hypothesis: **H_0**: $\mu = 500$ and **H_a**: $\mu = 500$. This is a **two-tailed curve** with the null hypothesis being rejected if the sample mean is too large or too small, that is, if the machine fills less than 500 containers **or** fills more than 500 containers.

- If the inventor had claimed that the machine filled at least 500 gallons, then the hypothesis is **H_0**: $\mu \geq 500$ and **H_a**: $\mu < 500$. This is a **one-tailed curve** with the rejection region on the left of the mean. The null hypothesis is rejected if the sample mean is too small. It is irrelevant to the hypothesis is the machine fills more than 500.

- Then again, if the inventor claimed that the machine filled **up to** 500 gallons, then **H_0**: $\mu \leq 500$ and **H_a** > 500. This is a **one-tailed curve** with the rejection region to the right of the mean. The null hypothesis is rejected if the sample mean is too large. In this case, it is relevant to the hypothesis if the machine fills more than 500 gallons but not if it fills less.

4 http://stattrek.com/Lesson5/HypothesisTesting.aspx?Tutorial=StatBased on Lesson 5; used with the permission of the author Harvey Berman.

STEP 2: FORMULATE AN ANALYSIS PLAN.

The analysis plan describes how you will test your data in order to accept or reject the null hypothesis. This decision is usually focused on a single test statistic. For interval data, the test statistic is one of: ***z-test*** with a mean for large sample; a ***t-test*** with a mean for a small sample; a ***t-test*** comparing the means of two independent samples; or a ***t-test*** comparing the means of dependent samples.

Other tests that may be used include the ***Pearson Product Moment Correlation*** test for relationships (interval data), the ***Spearman Rank Correlation*** for ordered (ordinal) data or the ***Chi-Square*** tests, one for goodness-of-fit for a proportion, the other to test for independence between two samples of nominal data. Note that each test uses a different formula. As part of the analysis plan, define the decision rule (a P-value or a rejection region) and determine the level of significance you will use. Generally, the level of significance used is 0.05 and 0.01, that is, 95% or 99%. Occasionally, if high precision is necessary, a significance level may be 0.001 (999%) or a low as 0.1 (10%).

Your analysis plan describes which tests you will apply to your data in order to accept or reject the null hypothesis. This unit discusses the basic procedures for interval data. Unit 4 discusses the tests listed in the previous paragraph. In discussing each test in Unit 4, criteria will be given to help you to know which test to choose.

STEP 3: ANALYZE SAMPLE DATA.

There are a number of tests that can be used to analyze your data. For example, mean score, proportion, t-score, z-score, etc. The one you are going to use should be named in your analysis plan. This section of the process is the part where you carry out the planned test and make the computations required. Separate units in this course describe each type of analysis and show examples of how to complete the test.

Find the value of the test statistic described in the analysis plan. Complete other computations, as required by the plan. In Unit 4, separate lessons describe each type of analysis and show examples of how to use an algorithm to complete the test.

Apply the decision rule. Your research project will be useless without interpreting the results of your research. The analysis plan should include how you will use the data you gather to accept or reject the null hypothesis. If the analysis of the data supports the null hypothesis, it is accepted; if not the null hypothesis is rejected. The terminology used is "fail to reject the null hypothesis."

STEP 4: INTERPRET RESULTS.

Your research project will be useless without interpreting the results of your research. The analysis plan should include how you will use the data you gather to accept or reject the null hypothesis. If the analysis of the data supports the null hypothesis, you fail to reject H_0, if not, then the null hypothesis is rejected. Part of the analysis will be to determine a ***calculated value*** for the statistic you use. This is then compared with a ***critical value*** read from an appropriate table. ***If the calculated value is less than the critical value, you reject the null hypothesis.*** But if you leave it at that, your colleagues will wonder what that means. You have to draw conclusions which go back to your research question. Your final effort will be to write the report that justifies the interpretation of your data and the conclusion(s) you come to as a result of your study.

The rest of this unit (3) describes what you need to do to prepare to analyze the data you will collect.

ANALYZING & INTERPRETING DATA:

There are six common tests that can be applied to research data. The simplest is the z-test. One attribute that distinguishes it from the t-test is that the z-test is usually used with large samples and the t-test with small samples. But there are other distinguishing characteristics. Other tests are Correlation tests and Chi-square tests. The rest of this unit (3) gives you the basic analyses that prepare you to do the other tests which are covered in Unit 4, Volume 2. Unit 4 will provide the information for you to analyze and interpret the results for your research study

BASIC ANALYTICAL TESTS:

The z-test requires tests of means as well as proportions. The means are averages. Proportions are ***ratios.*** Suppose, for example, that in a town of 600 people, 250 were females. Then the proportion is 250:600 or 25:60 or 5:12 reducing the proportion to its minimum.

The z-test is probably the easiest to understand. It analyzes means and proportions where there are one or two variables. Generally the z-test deals with large samples having 30 or more units of observation.

Sometimes professors are encouraged to grade on the curve. Analyzing the exam data using the z-test helps to grade the results on the curve. An example of this is shown in Unit 4.

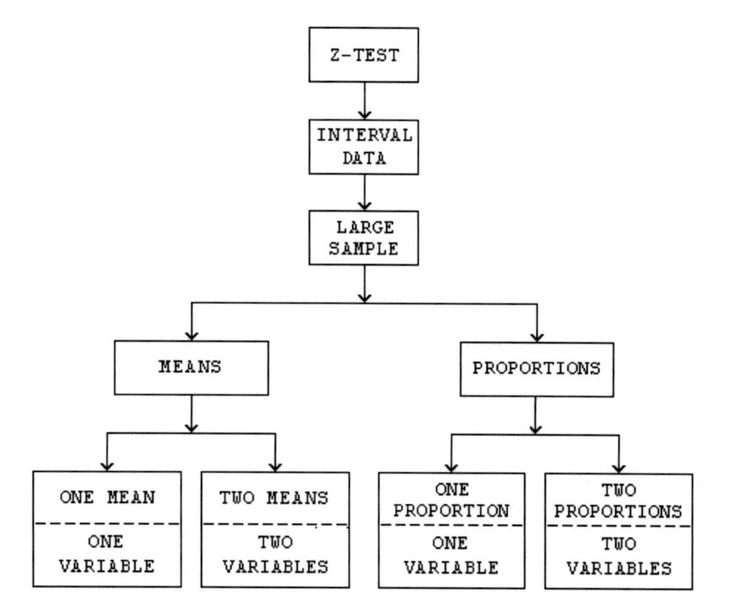

Figure 14: The z-test can be used to test means or proportions.

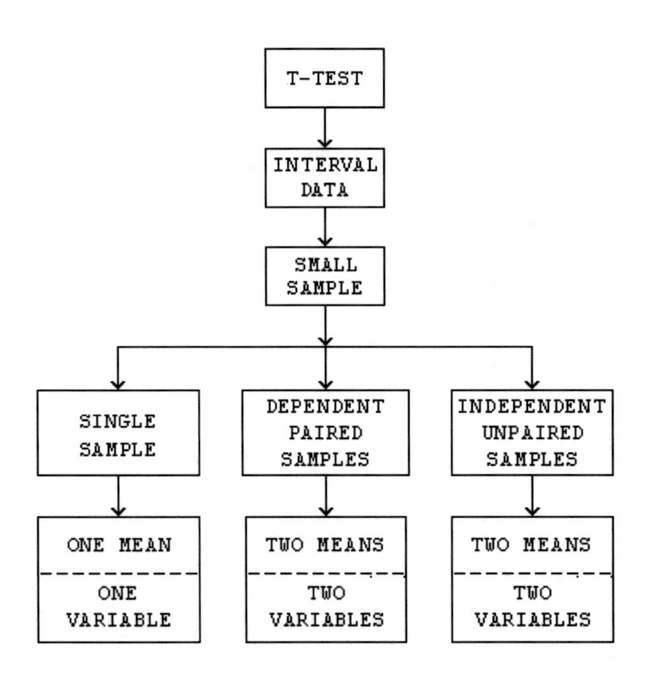

Figure 15: The t-test can be used to test a single sample or paired or unpaired samples.

The t-test is usually used with small samples. There is one type of distribution called ***The Student's t-Distribution***. The person who devised this was asked to name it, so he "named" the "owner" of the data "Student." It was an imaginary person.

An example of "paired and dependent" would be a professor comparing one group of his students at one time and the same group at either a different time, such as a pretest at the beginning of a semester and a posttest, the final at the end of the semester; or comparing the same students with a math test and a reading test. Since the group is the same for each test, they are paired and dependent.

An example of "independent and unpaired" samples would be two separate groups with no overlap between the two groups. In other words, the students in one group are independent of the students in the other group. The two groups may or may not have the same numbers of students. The only commonality between the two sections is probably the same topics of instruction and possibly the same instructor.

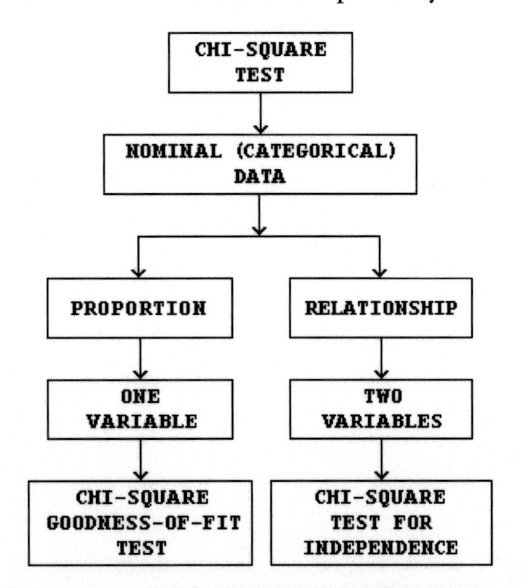

Figure 19: The Pearson and the Spearman correlations tests each test different types of data.

Correlation tests determine if there is any correlation or connection between the two sets of data. Note that each correlation test examines different types of data. The Pearson Correlation test is used on interval, that is, normal data. The Spearman test is used on ordinal data. Both test relationships, both have two variables for the comparison. Because the type of data tested is different, distinctive formulas are used and the results differ.

The Chi-Square test is used on nominal or categorical data. One is the goodness-of-fit test. It determines whether there is a significant difference between the ***expected*** value and the ***observed*** value. The observed values are your collected data, which you collected according to your theory as defined by your hypothesis.

The test for independence is used to analyze whether two different variables are associated with each other or are independent of each other.

One other use of the Chi-square test is to determine whether the population variance in a normal population has the same specified value as the variance of the sample.

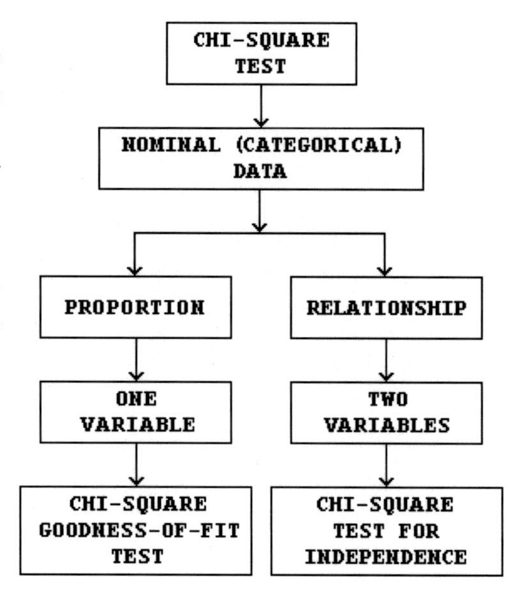

Figure 20: The Chi-Square test is used only for nominal data.

These tests call for tests of means or the variance or, although it has not been mentioned yet, the standard deviation discussed in the next Unit 3E.

UNIT 3A-2 Assignment:

1. Describe how the four steps in the process of testing hypotheses will be of value in **your** research project.

2. Describe the decision rules that enable **you** to accept or reject the null hypothesis.

3. Describe the differences between a one-tailed and two-tailed test including the identifying hypotheses.

4. As soon as you have collected all your data, take your hypothesis and complete the four steps for testing your hypothesis.

UNIT 3B: BASIC CONCEPTS FOR DATA ANALYSIS.

PURPOSE:

All statistical analysis is based on one or more principles of probability. For example, when a specific set of data is analyzed at a particular significance level or level of importance, it is usual to choose either 0.05, or 0.01, or 0.10. What these amounts mean are that, for example, at 0.05, the researcher claims there is a significance of 95%; or the probability that the researcher can reject the null hypothesis is at 95%, that there are only 5 chances in a hundred that the rejection is incorrect. A level of 0.01 means significance is at a 99% level and an 0.10 level gives a 90% probability. The choice of one of these levels of significance shows how confidant the researcher is in the results of their study. Clearly, the 99% level shows greatest confidence. Once you have gathered your data, your audience (those you need to convince that your study was worthwhile) will need to see your data represented in graphic form. Why? Most people will better understand your study if they can see, for example, a frequency distribution showing regions of acceptance and rejection. The lessons in this section have a number of different graphics which help to give examples of data under discussion in order to illustrate the concepts and make them more understandable.

OBJECTIVES:

- Describe the probability principles that are likely to affect the analysis of your data.
- Demonstrate how to represent a set of data in the form of a frequency distribution.
- Collect the data you need and use a frequency distribution for the interval data and an appropriate illustration or table to display the nominal and ordinal data (See APPENDIX A: Data Representation which shows the do's and don'ts of how to graphically represent different types of data).
- Illustrate **your** frequency distribution with a histogram, a frequency polygon and smooth curve.

DECISIONS RULES AND ERRORS:

Other types of probability; cumulative, conditional, and probability distributions. In addition to probability, there are a number of concepts that are included in this unit:

- Decision Rules.
- Decision Errors.
- Analysis Plans.

UNIT 3B-1: PROBABILITY.

PURPOSE:

The Unit 3A: *Preparing to Analyze your Data*, mentioned ***probability*** several times. Some definitions and explanation about probability is helpful.

A ***datapoint*** is a single item of data in your ***dataset***. Since your dataset is probably for a sample rather than the population, the datapoint may also be referred to as a ***sample point***. The collection of datapoints is referred to as a ***sample space***. All tests done on a sample space are dependent on the probability of something occurring. The following quote is from a series of Internet lessons on statistics[5] (emphasis added):

The ***probability*** of a sample point is a measure of the likelihood that the sample point will occur.

PROBABILITY OF A SAMPLE POINT:

By convention, statisticians have agreed on the following rules.

1. The probability of any **sample point** occurring can range from 0 to 1.
2. The sum of probabilities of all **sample points** in a **sample space** is equal to 1.

OBJECTIVE:

- Describe the probability principles that are likely to affect the analysis of your data.

5 http://stattrek.com/Lesson2/ProbabilityDistribution.aspx
Source: "Tutorial: Probability Distributions." Information is used with the permission of the author Harvey Berman.

INTRODUCTION:

Some sayings about statistics that were sent recently in an e-mail (anonymous):

1. 42.7 percent of all statistics are made up on the spot.
2. Remember, half the people you know are below average.
3. 99 percent of lawyers give the rest a bad name.
4. Statistics makes tiny difference you can barely see into something significant.

Research studies which provide interval data are usually some form of an experiment. The outcomes of the experiment, the data, are often referred to as events. The whole set of data, whether outcomes or events, are a ***sample space*** or a ***dataset*** and are analyzed using rules of probability. The following definitions are needed for understanding probability as it applies to statistical research.

VARIABLE: A variable is any usually alphabetic symbol that represents a set of quantitative or numeric values (e.g., X, Y, a, b, n, etc.).

RANDOM VARIABLE: The set of variables that are the outcomes found in the study of a particular research question.

X: ***X*** represents a set of observations and may be written:

$X_1 + X_2 + X_3 + \ldots + X_n$ meaning the first, second, third, etc., to the nth value found, where there are a total of ***n*** observations.

X_1, X_2: When two sets of data are collected for the same group at two different times (such as exam scores), the first set of scores is represented by X_1, and the second set is represented by X_2.

X, Y: If there are two different variables being studied (such as exam score for two different sets of students in the same topic, no overlap of students), one set is represented by ***X*** and the other by ***Y***.

N: The "unknown" value shown by ***n*** represents the number of observations collected. It always represents one of the natural numbers: 1, 2, 3, etc.

N_1, N_2: When two variable are collected, n_1 and n_2 represent the number of observations in the two data sets X_1, X_2 OR ***X, Y***. Depending on circumstances, n_1 may be equal to n_2 but it may not.

OUTCOME: The anticipated result of a statistical experiment given in the form of a probability statement such as $P(X = x)$ meaning the probability of ***X*** being equal to ***x***.

PROBABILITY OF X: $P(X)$ represents the probability of ***X*** occurring. $P(X = x)$ represents the probability of ***X*** having a particular value such as ***x*** (lower case), for example If $x = 1$, then $P(X = 1)$.

SET: A collection of ***elements***, for example, the set of the first five natural or counting numbers is {1, 2, 3, 4, 5}. The numbers in this set are the elements of the set. When gathering data, the elements of the set may be referred to as ***observations*** or ***datapoints***.

SAMPLE SPACE OR DATASET: The sample space is the whole collection of observations for the variable being considered.

PROBABILITY & PROBABILITY DISTRIBUTIONS:

Probability is a way of measuring the likelihood of a particular experimental outcome occurring, in other words, ***estimating*** the likely occurrence of some event. For example, the table below shows the probabilities associated with the outcomes of tossing a coin two times to get H (heads) or T (tails). There are four possible outcomes:

TT both tosses result in tails, that is, zero heads,

HT a head followed by a tail, that is, one head,

TH a tail followed by a head, that is, one head,

HH both tosses result in heads, that is, two heads.

The statements "head followed by tail" and "tail followed by head" are unique as stated. However, by summarizing each statement as "one head," the results are no longer unique.

The probability of any one of the above outcomes occurring is 1 in 4 or 1/4 or 0.25. These values assume that each outcome is unique. But two of the outcomes, HT and TH give the same result if we are considering only the probability of a head resulting from the coin toss rather than the sequence of tail and head. If we let ***X*** represent the number of heads that are the possible outcomes, then the probability is that values of ***X*** are 0, 1, or 2.

Since there are two ways a single head can result, the possibility of getting one head is twice times 0.25 or 0.50 as shown in Figure 21.

Number of Heads	Probability
0	0.25
1	0.50
2	0.25

Figure 21: The probability of occurrence is based on the number of heads.

In advance of collecting the data, the possible outcomes are theoretical, use probabilities, and are not based on observations. What the above probability distribution is saying is that ***theoretically,*** if you toss two coins together 100 times, 25 tosses will produce tails, 50 tosses will produce one head and 25 tosses will produce two heads. However, if you perform the coin toss ***experiment***, the ***actual results,*** the observations, of tossing the two coins together may differ because ***each time you toss the coin the probability of getting one head is exactly the same.*** If you toss the coins one hundred times, you may get a single head much more often than 50 times or tails may turn up much more often than 25% of the time.

Another example in probability distributions: In a laboratory, the researcher has noticed that each of his experimental mice has not less than three offspring in a single birth. Now he is looking for information on the possible gender of the first three offspring of a particular animal, either M (male) or (F) female. One ***possible*** outcome is that all three of the offspring will be M (male), a prediction based on the probability of occurrence ***before*** the births take place.

Figure 22 is a ***sample space*** with a total of eight (8) ***possible outcomes*** (that is theoretically) for the first three offspring with each offspring's birth being an event:

EVENT	EVENT	EVENT	OUTCOME	OUTCOME DESCRIPTION
M	M	M	MMM	All three are male.
M	M	F	MMF	Two males followed by one female.
M	F	M	MFM	One male, then one female, then one male.
M	F	F	MFF	One male followed by two females.
F	M	M	FMM	One female followed by two males.
F	M	F	FMF	One female, then one male, then one female.
F	F	M	FFM	Two females followed by one male.
F	F	F	FFF	All three are female.

Figure 22: The probability outcomes of events in the birth of the first three offspring.

There is an equal probability that a male or a female will be born in the first event. That is 50% or 0.5. Each event has the same probability including the third event (0.5 each time). The probability of three males occurring, MMM, is the ***product*** of the three probabilities: 0.5 x 0.5 x 0.5 or 0.125.

A TREE DIAGRAM:

Three males, MMM, is the outcome of the first set of branches in the Tree Diagram. Each outcome (end of each branch of the tree, Figure 22) has a one in eight chance of occurring or 1/8 or 0.125. See tree diagram on the next page.

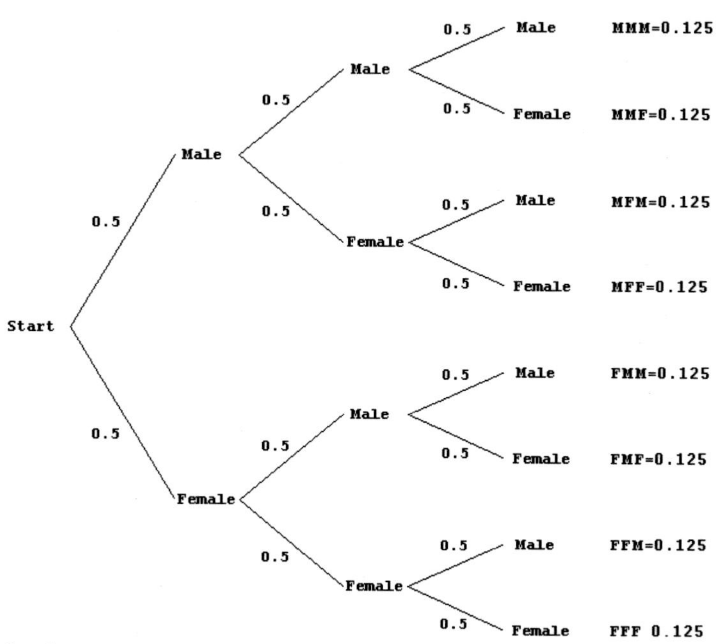

Figure 16: A tree diagram showing probabilities of each of eight outcomes and also the first three offspring in each branch.

WHAT IS A CUMULATIVE PROBABILITY?

A probability distribution gives the chance of any one of the events (outcomes) occurring in isolation. Notice that the probabilities of each total event in the tree diagram are 0.125 and they sum to one (1).If we were interested in the likelihood of two males followed by a female, there is only one outcome, ***M M F.*** On the other hand, if we were interested in the possibility of having two males and one female regardless of order, there are three outcomes: ***M M F; M F M;*** and ***F M M.*** Probability is 3 chances in 8 or 0.375. In addition you can calculate the probability by adding each ***P(X)*** together, 0.125 + 0.125 + 0.125 = 0.375. This is known as a ***cumulative probability.***

EVENT X	P(X)
MMM	0.125
MMF	0.125
MFM	0.125
MFF	0.125
FMM	0.125
FMF	0.125
FFM	0.125
FFF	0.125

OUTCOME DESCRIPTION	
All three are male.	
Two males followed by one female.	✓
One male, then one female, then one male.	✓
One male followed by two females.	✓
One female followed by two males.	
One female, then one male, then one female.	
Two females followed by one male.	
All three are female.	

Figure 24: Three of the outcomes (checked) fit the requirements for the cumulative probability.

A cumulative probability is a ***sum of probabilities*** and refers to the probability that the value of a random variable (such as ***X***) falls within a specified ***range.*** For example, suppose you flipped a coin two times. The probability of the third flip producing one head or less is the cumulative probability of the coin showing no heads and the coin showing one head which is 0.75 or 75% (see table below). The probability for the middle column is a simple probability that ***P(X = x).*** The variables are ***X*** (capital letter) which is a random variable, and ***x*** (lowercase letter) which is one of ***X***'s values. The probability of the coin showing ***one or less*** heads in two tosses is ***P(X ≤ x)*** or 0.75.[6]

Figure 25: An empirical test provides this table of probabilities.

No of Spots	Probability
1	0.10
2	0.25
3	0.20
4	0.15
5	0.10
6	0.20
Total	1.00

6 http://stattrek.com/Lesson2/ProbabilityDistribution.aspx
Tutorial: Probability Distributions. Used with the permission of Harvey Berman, author.

RELATIVE FREQUENCY:

When a die is rolled, any one of six sides may turn up. It is customary to assign probabilities of one in six or 1/6 (one chance in six rolls) to each side which assumes that each side has an equal chance of showing. Since we don't know whether the die is perfectly balanced or not, we cannot ***assume*** that each side has an equal chance of showing up. By rolling the die a large number of times we can record the ***frequency*** with which each side turns up. The frequency is used for further calculation. Let's suppose a die has been tested by tossing it 100 times. The measurements have been converted to the probabilities shown in the table. Remember: the total probability must be equal to one (1).

EVENT X	P(X)
MMM	0.125
MMF	0.125
MFM	0.125
MFF	0.125
FMM	0.125
FMF	0.125
FFM	0.125
FFF	0.125

OUTCOME DESCRIPTION
All three are male.
Two males followed by one female.
One male, then one female, then one male.
One male followed by two females.
One female followed by two males.
One female, then one male, then one female.
Two females followed by one male.
All three are female.

Figure 26: The first condition, shown by the white area is for the firstborn to be male. A possible alternate condition, shown by the highlighted area, could be for the firstborn to be female.

The ***relative frequency*** becomes the relationship between the probability of any one side turning up and the total probability. This is an empirical approach, based on or verifiable by observation or experience rather than theory or pure logic.

If you wanted to know the relative probability of a 6 showing, it is 10/1.00 or 10, since the total probability has to be 1.00. If you wanted to know the relative frequency of ***either*** a 2 or a 3 turning up, it is a ***cumulative probability*** that equals 0.45.

CONDITIONAL PROBABILITY:

Sometimes, two events are connected, for example, when event ***B*** cannot occur unless event ***A*** occurs first. That is, the likelihood of event ***B*** occurring is ***conditional*** on the occurrence of event ***A***.

EVENT A	EVENT X	EVENT B
M	MMM	–
M	MMF	MF
M	MFM	FM
M	MFF	FF

Figure 27: Eliminate the first event because there are no female offspring, a required condition.

Suppose the researcher wants to find the probability of his animal having at least one female if her first offspring is male? In Figure 25, only the first four events have a male for the first of three offspring. However, first, the other four events have a female as the first offspring, which is not the required condition. Let's extract the first four events, below. The condition given is that the first offspring is male. There are four outcomes that fit this condition, but there is a second condition which requires there to be at least one female following the male.

Note that this is not a cumulative probability. The three outcomes fulfilling the requirement are outcomes that are ***conditional*** on the first offspring being male and that there is at least one female following. The conditional probability is 3 of 4 occurrences, or 0.75.

Now, the requirements of "conditional probability" include the following: If the whole sample space is considered, the probability of the first event being a male is 4 out of 8 outcomes or 0.5. But then the probability of having at least one female following the male is a ***cumulative*** not a ***conditional*** probability of 3 out of 8 or 0.375. Why? Because if you use the whole sample space you are placing no conditions on the events you are looking at.

UNIT 3B-1 Assignment:

1. Discuss the various types of probability in your discussion group, then in your notebook describe in your own words the following probability principles that are likely to affect the analysis of your data.

    a. Probability as an outcome in a sample space.

    b. Probability distributions.

    c. Cumulative probability.

    d. Relative frequency.

    e. Subjective probability.

    f. Conditional probability.

2. Draw a tree diagram with three possible outcomes; for example, Rain, Cloudy, and Sunny.

UNIT 3B-2: FREQUENCY DISTRIBUTIONS

PURPOSE:

Don't forget, doing research is done an effort to not only educate yourself but also to enlighten others through your report. Using illustrations in the text of your report will enable future readers to more clearly understand what you were trying to do, how you did it, and your conclusions from the results. It is usually appropriate to include an appendix with examples of your data collection methods, materials used in your study (photos or actual examples), data arrays, how you arrived at your conclusions, and whatever helps to clarify your intent.

When you have collected ***all*** your data, then you can use arrays, frequency tables, or frequency distributions, etc. This unit shows a number of ways to record frequencies.

OBJECTIVES:

- Demonstrate how to represent a set of data in the form of a frequency distribution.
- Collect the data you need and use a frequency distribution for the interval data and an appropriate illustration or table to display the nominal and ordinal data (See APPENDIX A: Data Representation which shows how to graphically represent different types of data).
- Illustrate **your** frequency distribution with a histogram, a frequency polygon and smooth curve.

INTRODUCTION:

In your study, you were required to get all three types of data so you have the experience of actually working with each type. An empty copy of the instrument you used in gathering data can be used as a score sheet. You may have already collected the data needed to test your hypothesis, If you are still in the process, just keep going. When you have obtained your data, you will need to display it in a form that can be easily interpreted. The simplest method for numeric (interval) data is a ***frequency distribution.***

THE FREQUENCY DISTRIBUTION AND FREQUENCY TABLE:

The data you have collected is referred to as raw data before you do anything to it. Data may be on a continuum, such as age or height, or discrete, that is, individually distinct, such as the number of siblings or the number of times each member of a team scores in a game. In the raw state, the data is at first unsorted. Until it is sorted, there is little information that can be derived from the dataset.

Here is ***raw data***, displayed in an array, concerning the ages of 70 students enrolled in an undergraduate History course.

UNSORTED ARRAY:

18, 21, 23, 22, 17, 25, 27, 22, 27, 19, 20, 28, 18, 19,
21, 20, 26, 25, 19, 24, 18, 17, 21, 20, 21, 18, 19, 21,
21, 25, 20, 22, 23, 19, 20, 21, 20, 26, 18, 19, 20, 25,
19, 21, 20, 17, 22, 20, 18, 19, 26, 22, 20, 22, 20, 21,
19, 18, 17, 24, 25, 19, 19, 28, 26, 23, 23, 24, 22, 24

Tabulation refers to the arrangement of data in columns or tables or in an array. ***Frequency*** relates to how often a particular event or data-point occurs. When the data is sorted, it is much easier to determine the frequencies. You can use one of two methods to sort the data:

1. Sort the numbers themselves, or

2. Complete a tally sheet.

Both methods are demonstrated here. The array is sorted (by hand or computer). The sorted array allows you to see each data-point of the ages so you can count the number in each age classification.

SORTED ARRAY:

This array is sorted in ascending order, that is from lowest to highest, which is probably the commonest method. Depending on the type of information you have, it is acceptable to sort in descending order (highest to lowest). Although this array is sorted it still contains raw data.

17, 17, 17, 17, 18, 18, 18, 18, 18, 18, 18, 19, 19, 19,
19, 19, 19, 19, 19, 19, 19, 19, 20, 20, 20, 20, 20, 20,
20, 20, 20, 20, 20, 20, 21, 21, 21, 21, 21, 21, 21, 21,
21, 21, 21, 21, 21, 22, 22, 22, 22, 22, 22, 22, 22, 22,
22, 23, 23, 23, 23, 23, 23, 23, 23, 25, 25, 25, 25, 25

A FREQUENCY ARRAY:

When you start to count the number of items in a particular class of data, such as the number of 17-year olds, the frequencies that result are no longer considered to be raw data. At this point, the data can be placed in a ***frequency table***: Counting the totals in the array: 4 x 17; 7 x 18; 11 x 19; 12 x 20; 13 x 21; 10 x 22; 8 x 23; 0 x 24; 5 x 25.

Notice that the age 24 has a frequency of zero. It still must be included as part of the data as it is just a matter of chance that there are no 24-year-olds in this particular dataset.

AGE	FREQUENCY
17	4
18	7
19	11
20	12
21	13
22	10
23	8
24	0
25	5
Total	70

Figure 28: The frequency array summarizes the data in the sorted array.

THE TALLY SHEET:

Age	Tally	Total
17	IIII	4
18	~~IIII~~ II	7
19	~~IIII~~ ~~IIII~~ I	11
20	~~IIII~~ ~~IIII~~ II	12
21	~~IIII~~ ~~IIII~~ III	13
22	~~IIII~~ ~~IIII~~	10
23	~~IIII~~ I	8
24		0
25	~~IIII~~	5
	Total	70

Figure 29: Record the frequency of occurrence with slashes.

The tally sheet allows you to make a single vertical mark or stroke for each score. To help make the tally easy to read, after marking four strokes, the fifth is an angled line drawn across the first four from top left to bottom right. When counting, you count the groups of five first, then add any single strokes to the total. Here is an example of a tally sheet using the same data as the sorted array.

In some ways, the tally sheet is preferable as the information can be seen at a glance. One problem in using tally sheets is that you have to be very careful to ensure ***all*** data are tallied. Make a copy of your unsorted data and mark off each data-point as you tally it.

CLASS, CLASS INTERVALS, AND FREQUENCIES:

On the tally sheet each different age is referred to as a class, related to the word "classification," in other words, each class is a category of items that have some property in common. In the above example, the property is age, and all the people in each class have a specific age in common.

Range:

The range of a set of data is the distance between certain limits or the difference between the smallest and the largest category. Examining either the sorted array or the tally sheet, it is clear that the range is 25-17 = 8. So the range of ages in the History course is 8.

Interval:

An interval occurs when data is grouped, for example, the above data could be grouped by ages such as 15-19, 20-24, 25-29. An "interval" may be an interval of time or space. The boundaries of each interval are the upper and lower limits of the interval. The values between 15 and 19 are an interval of 5 bounded by 15 as the lowest value and 19 as the highest.

Class Interval:

A range of numbers defined arbitrarily by the highest and lowest numbers in the class. So for the class of 40-49, the class interval is the range of numbers 40 through 49.

Suppose we were looking at the test scores of the same 70 students. Putting the scores into intervals works well. Instead of a class of a single age, we now have a class of scores from 40 to 49 in the first interval with a range of 10 numbers. There are six intervals with the range of each interval being 10.

SCORES	FREQUENCY
40-49	3
50-59	5
60-69	25
70-79	22
80-89	10
90-99	5
Total	70

Figure 30: A frequency table using class interval data.

One might ask: "Why use 40-49, etc., why not use 41-50?"

When one has large volumes of data, it is simpler if the computer crunches the numbers. It is easier to have a program sort the data by the first digit of the number rather than both digits. For example, with the class intervals of 40-49, 50-59, 60-69, 70-79, etc., the computer can gather all the numbers beginning with 4 into one group, the numbers beginning with 5 in a second, the numbers beginning with 6 in a third, etc. Compare that with class intervals of 41-50, 51-60, 61-70 etc. With the latter, you cannot use the first digit for the sorting. So, why not keep it simple?

Now if the range of our data-set begins with 0, how do we treat the interval 0-9 as there is only a single digit? This is also easy: Make it a double-digit by placing a 0 (zero) in front of the single digit, 00, 01, 02, etc. Caution: If you have data in the 0-9 range and use only single digits, the computer might include 3s with the 30s; 6s with the 60s; 9s with the 90s, and so on.

It is possible to make unequal class intervals for a particular set of data. For example, when comparing income, the intervals might be: below 15,000; 15,000-30,000; 30,000-60,000; 60,000-100,000 etc. The first two intervals have a range of 15,000 each; the third interval has a range of 30,000 and the fourth has a range of 40,000. For simplicity in your studies, it is recommended that equal-sized intervals be used. If a researcher were to use unequal-sized intervals, he or she would need a very good reason to justify it.

FREQUENCY OF A CLASS INTERVAL:

The frequency is the number of times a particular value or event is observed. So the frequency is the number of values that fall within a particular class interval. For example, the number of scores in the 60-69 class interval is 25, so the frequency of that class interval is 25.

RELATIVE FREQUENCY:

This refers to the relationship between the frequency ***within*** an interval and the ***total number of measurements made.*** So the frequency of the class interval of 60-69 compared to the total number of measurements is the relative frequency of 25:70 or 35.7%. The relative frequency of the 80-89 class interval is 10:70 or 14.3%.

FREQUENCY TABLE & FREQUENCY DISTRIBUTION:

A systematic arrangement of values grouped into class intervals, used to summarize data so that the frequency of each interval is clearly displayed, and the relative frequency can be easily calculated. Suppose the same 70 students were to take another test. Their scores are the frequency in each interval and the quantity of those scores is graphed as a frequency distribution. Here are two methods of graphing data distributions: Intervals for exam scores in Figure 2 have a range of 10 each, the number of class intervals of interest is 6. In a bar graph, on the right below, the bars are separated by a small space. In a ***histogram*** the "bars" are touching. The line graph is NOT a curve; the lines touch the plotted points at angles. It could be the basis for a curve.

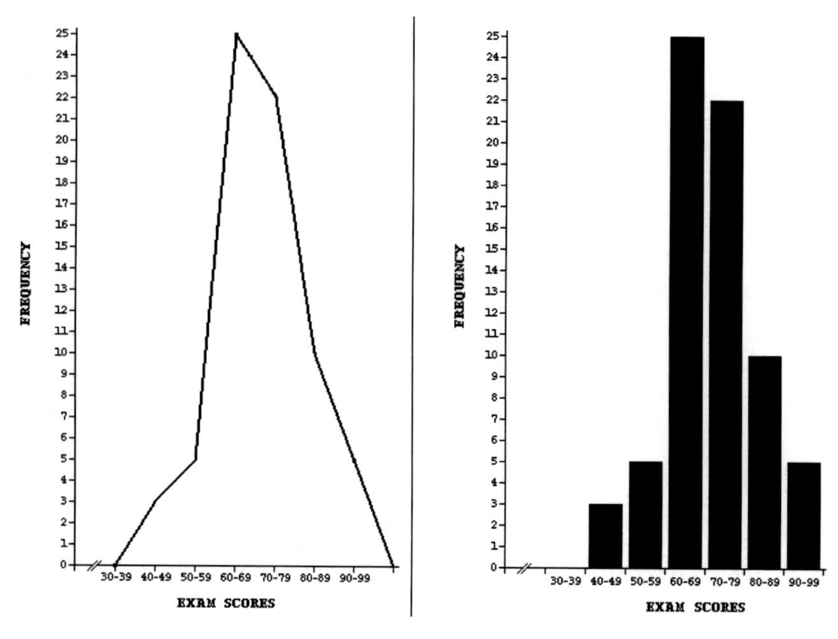

Figure 31: Two methods of illustrating the same data: a line graph and a histogram.

These two graphs were intentionally drawn this tall, even though it meant smaller numbers, to illustrate a point: It was necessary because of (1) the scale needed to show the exam scores and (2) the scale of the frequencies.

The two graphs were combined in a single illustration so the two different styles could be compared.

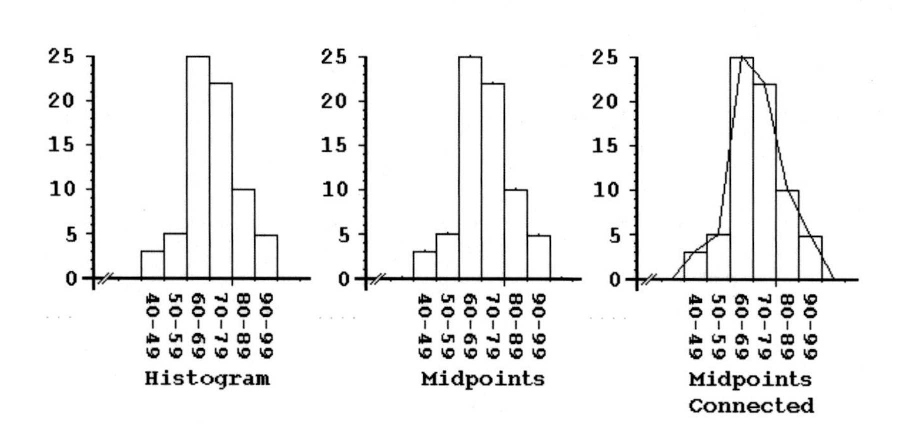

Frequency is from 0 (zero) to 25; exam scores are intervals from 30-39 through 90-99. The histogram is based on the same data but adjustments to the intervals for the scores had to be made to facilitate the placement of the bars.

Figure 32: The first three stages for preparing to convert a histogram to a smooth curve: a normal histogram, a histogram using the midpoints of each class interval, and a histogram with the midpoints connected in a frequency polygon.

One way to solve the size problem would be to sort the frequencies into class intervals; however, that would cause a problem in showing the range of the data between zero and four, or zero and nine depending on which range was chosen. The frequency range could be two: zero to one, two to three, etc. This illustration is shown as is to encourage you to be careful in choosing how to illustrate your data.

USING CLASS INTERVALS:

Suppose you were surveying all the people in a small town with approximately 1,200 people. Your data concerning the ages of the people in town shows a range from a few days through 99 years. There are no centenarians. A useful interval would be either five years or ten, depending on what you need to know; whichever you used, the class intervals would start with 0, with newborns and children a few months old being classed as zero years old. The five year intervals would be 00-04, 05-09, 10-14, 15-19, and so on. Ten year intervals would be 00-09, 10-19, 20-29, etc. Which range of intervals you chose would depend on what you were trying to determine.

The larger interval increases the frequency within each class interval but reduces the number of class intervals being considered. An alternative: by grouping the people in town into "families" that allows for couples and singles if living on their own, you would reduce the survey forms needed from 1,200 to 300 assuming an average of four per family; or 400 assuming an average of three per family.

When using class intervals, the midpoint of each interval is important. The midpoint of an uneven quantity is easy to determine. For example, in the 20-24 class interval the midpoint is 22. In the 60-69 class interval the midpoint is between 65 and 66. To calculate it, take the upper boundary (69) and subtract the lower boundary (60) and divide by 2 or 9÷2 = 4.5 then since the class interval is 60-69 (not 00-09) add the result to the lower boundary. The midpoint is 60 + 4.5 = 64.5. When working with numbers, rounding may be required. For example: anything between 60.0 and 64.4 rounds down (60-64) and anything between 60.5 and 68.9 rounds up (65-69).

The Frequency Histogram:

You will need to illustrate your data. A number of illustrations are shown here to give you several ideas. One graphic illustration of data with class intervals is the ***histogram;*** the diagram consists of rectangles. Class intervals are shown by the ***width*** of the rectangle, and the ***height*** of the rectangle represents the frequency.

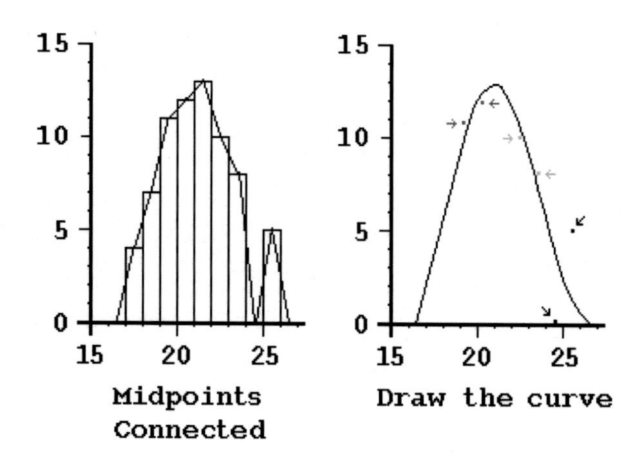

Figure 33: The final stages of drawing a (comparatively) smooth curve. You may need to "average" data in some places as shown by matching arrows.

If your data has a single number for each class, the width of the rectangle is one. With class intervals of five or ten, the width can be adjusted as preferred. The scale along the horizontal axis defines the ***categories of age.*** The rectangle is usually drawn with the left edge at the value being defined by the rectangle so that the 40-49 interval starts at the value 40.

Midpoints:

The midpoints are drawn at the center point of each rectangle ***and at the midpoint of the unused frequency on either end of the data.*** Connecting the midpoints includes any midpoints representing zero frequencies. The frequency polygon is the shape defined by connecting midpoints with straight lines. Close the polygon at each end by meeting the horizontal axis at mid-point of the hypothetical adjacent classes below left and right. The midpoints are joined to show the broad outline for the curve. See Figure 32 for the final stages of drawing the smooth curve.

Smooth Curve:

Researchers usually prefer to see data drawn in a curve. Where the curve cannot go through midpoints, it is drawn equidistant from the adjacent midpoints (see matching arrows).

In this example, if a vertical line were to be drawn from the peak of the curve to the baseline, the two parts of the curve on either side of the line would not match. The reason: the data is slightly skewed to the right. The skew is caused by the empty datapoint and the datapoint to the right of that.

NORMAL DISTRIBUTION:

Many natural phenomena fit this type of data distribution. The central point of a ***normal curve*** (shown on the left below)is exactly in the center, drawn vertically, and the left and right halves of the curve are exactly equal in area with one being the mirror image of the other. The normal or bell curve is ***symmetrical*** around the vertical axis. Many times a curve is skewed left or right, when skewed to the right, it is positive (sometimes written as: +ve) and has a "tail" pointing to the right; skewed to the left it is negative (sometimes written as: -ve) with the "tail" pointing to the left. Data which causes a skew in the curve is usually referred to as ***outliers.*** If your data shows one or more outliers it may be necessary to make allowances. Generally skewed data is not desirable when trying to determine if your null hypothesis can be rejected or not.

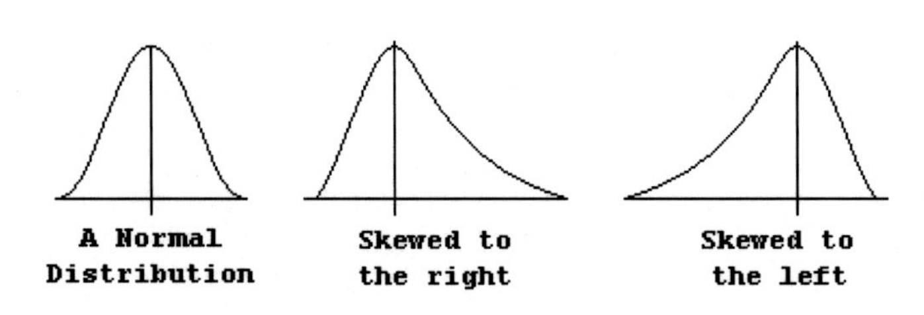

Figure 34: Comparing a normal distribution with skewed curves.

The skewed ends of the curves are caused by "outliers" or data that lie outside the normal range of a curve.

AN EXAMPLE OF POTENTIALLY SKEWED DATA:

The Professor of Astronomy 101 had a large class of undergraduates. One person was taking the class because it was a requirement for her major; she had been very interested in astronomy since junior high. Her scores in the final exam gave her top marks and the student who came in second had 17 points fewer. Her score was unusual, and is what is known as a ***gross outlier*** because it lies well outside the normal range of data.

Now, if the Professor had included her score as part of the final evaluation the grading curve would have terribly skewed as the second student would have only received a B+. Instead, the professor gave her an A+ for the semester. He then ignored her score and acted as if the next student came top of the class. That way, that student was given an A (which he did deserve) and the rest of the students fit the normal curve.

UNIT 3B-2 Assignment:

Your plans for your study should be at the point where you can collect and begin to organize your data. As it may take some time to actually collect the data, while you are working on Unit 4: Procedures for Analyzing Data, do whatever is necessary to get your data organized. The following assignment will be completed after you have collected all of your data.

1. Sort your data and use a frequency table for the interval data, and an appropriate illustration or table to display the nominal and ordinal data.

2. Illustrate the frequency distribution with a histogram, a frequency polygon, and a smooth curve (model them on the graphics in this Unit).

3. Write a report of how you collected your data and the results, illustrating the information appropriately. Include what you learned and any mistakes you had to correct or problems you had to solve.

4. Critique the reports of your fellow students and discuss the reports with a view to improving them for when you write your final reports.

5. Hand in the original and improved report and the critiques.

UNIT 3C: PRELIMINARIES FOR TESTING INTERVAL DATA.

PURPOSE:

Research is the science of collecting, classifying, tabulating and testing data so that significant information can be presented about a given subject. At this point, you should have begun collecting your data. In this unit, you will finish collecting data, complete your preparations and begin the process of testing your hypothesis. There will be plenty of practice activities for each test you learn and NONE of the calculations will be complicated. Algorithms are included giving a step-by-step process for each test. Just follow the steps carefully and you should have no problems.

OBJECTIVES:

- Describe the value of determining the mean and standard deviation to a set of Interval Data.
- When you have finished collecting your own data, calculate the mean and standard deviation.
- Explain how the mean and standard deviation provide a good foundation for more complex calculations.

FLOW CHART - ANALYZING INTERVAL DATA:

Interval data is the commonest type of data collected in a research study. Before you can use the various tests, you need to be able to calculate the mean, variance and standard deviation. Section 3C-1 discusses how to calculate the mean, median and mode, the three Measures of Central Tendency. Section 3C-2 discusses the Measures of Spread, range, variance and standard deviation. Analysis of both z-test and t-test requires not only the mean but also the standard deviation. The Pearson correlation test is not discussed until Section 4 in the second volume of this book.

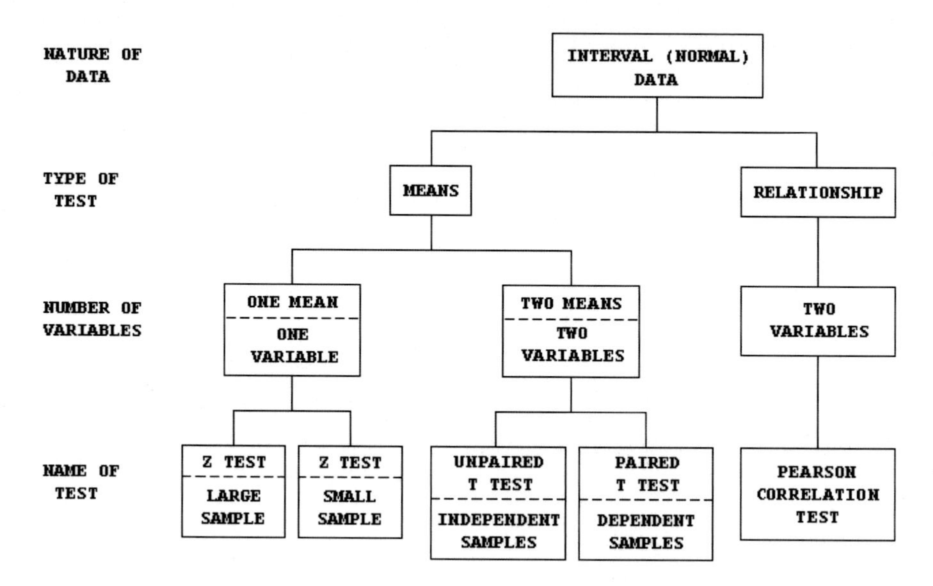

Figure 35: The z-test, the t-test and the Pearson Correlation test are the three tests that may be used on Interval data.

3C-1: MEASURES OF CENTRAL TENDENCY - MEAN, MEDIAN & MODE

PURPOSE:

As discussed in Unit 3B. *Basic Concepts for Data Analysis*, before you begin testing your interval data you must sort it into ***frequency distributions.*** Once you have a frequency distribution, you are able to find the measures of ***central tendency***, the ***mean, median,*** and the ***mode.***

The mean allows you to calculate the ***variance.*** Once you have the variance, it is a simple matter to determine the ***standard deviation.*** You will not be required to do complicated analyses at any time in this semester.

OBJECTIVES:

- Describe the difference between the terms: mean, median, and mode and how each one is calculated.
- Given data in the form of a frequency distribution, calculate the mean, median and mode.
- Calculate the mean, median, and mode for **your** frequency distribution.

INTRODUCTION:

The most likely data you will collect in *any* research study is interval data. You will probably also collect either nominal or ordinal data or both. For your current study, you are required to collect all three, to gain experience. Since the commonest type of data collected is interval data, the most useful place to start analyzing such data is the average or ***mean.*** The mean is calculated on the frequency of observations and results in an average for the dataset.

"I CAN READ YOUR MIND!"

A game played by children requires you to calculate a mean. The game goes like this, to the great mystification of those who don't know the secret. The use of the words "double it" and "divide by two" are deliberately misleading as they imply they are totally different actions instead of the second reversing the action of the first.

- Think of a number between one and ten; don't tell me what it is ***(e.g., 7).***
- Double it ***(14).***
- Add 8 ***(note, always add an even number; 14 + 8 = 22).***
- Divide by two ***(you now have the average of 22/2 = 11).***
- Take away the number you first thought of ***(11 - 7 = 4).***
- The answer is 4 ***(half the value you added in step 3).***

Step 4 calculates the average or mean between two values (14 and 8). Since there are only two values (see step 3), the sum is divided by 2. Because the original number is subtracted in step 5, you never need to know what it was.

THE MEAN, MEDIAN, AND MODE:

The three measures help define the curve. These are considered to be the most representative measurement(s) in a set of data. In a normal distribution, mean, median, and mode are equal in value and are defined by the central vertical axis.

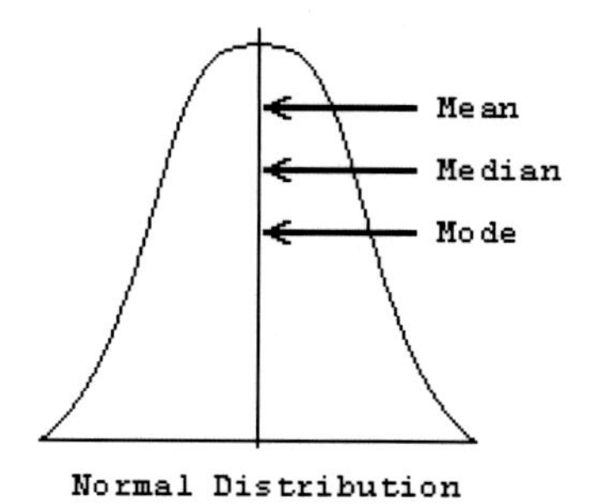

Figure 36: The even curve of a normal distribution ensures that the mean, median and mode are all equal.

It is conventional, when drawing a curve to represent the frequency, to have the ***lowest value on the left*** with the others falling on a continuum to the right with the highest values at the right end of the curve.

A lot of data have normal distributions. It is when we have data that does not have a normal distribution that the mean, median and mode give us additional information about our data by helping to define the shape of the distribution. Let's look at each of these measures of Central Tendency.

THE MEAN:

The mean is the arithmetic average of the combined set of data. It is found by dividing the ***sum*** of all measurements by the ***number*** of measurement in the set. One caution: the average can be distorted by extreme values. For example, suppose a number of measurements were gathered and all but one fell between 30 and 50; the one measurement was 3. Including that data point would skew the average to the left. The proper name for these data points is ***outliers***, in the sense that they lie outside the "normal" spread or range of data.. In the case of an extreme datapoint far outside the normal range, they are referred is gross outliers.

Since they tend to distort final results, the temptation is to discard them. But before deciding to eliminate them, remember that if you don't include them, your data-set is incomplete. One way to include them is to use class intervals with the 3 in the above example being included in the bottom class interval by calling it "less than 30." If in the same data set there had been a datapoint of, for example, 75, that could be included in a class interval of "greater than 50."

CALCULATING THE MEAN:

EXAMPLE 1:

Suppose there are 23 measurements in your data-set.

The "raw" (unchanged) data:

1, 4, 3, 5, 5, 1, 7, 8, 3, 7, 1, 2, 7, 2, 5, 6, 1, 3, 8, 7, 3, 2, 1

It is usually easier to deal with your data after it is ordered:

1, 1, 1, 1, 1, 2, 2, 2, 3, 3, 3, 3, 4, 5, 5, 5, 6, 7, 7, 7, 7, 8, 8

Tallied Scores:

X	TALLY	TOTALS
1	~~IIII~~	5
2	III	3
3	IIII	4
4	I	1
5	III	3
6	I	1
7	IIII	4
8	II	2
		n = 23

Figure 37: "Tally" is the process for finding the frequency of each occurrence.

X contains the numbers 1, 2, 3, 4, 5, 6, 7, 8. These are ***labels*** not interval data. The number may represent inches, hours, a quantity of candy, or some other fixed item. What we are tallying is the number of times each label occurs. The number represented by the tally is the ***frequency*** of the occurrence of each label.

FREQUENCIES:

X	FREQUENCY
1	5
2	3
3	4
4	1
5	3
6	1
7	4
8	2
Total	n = 23

Figure 38: A frequency table.

X contains the numbers (labels) 1, 2, 3, 4, 5, 6, 7, 8.

The frequency of each ***X*** is shown in the second column. The frequency, such as 5, shows the number of instances in which the variable ***X*** takes on its possible value of "1"; ***X*** is a generic term representing each of the measurements that can be made.

The total number of measurements ***n*** = 23; ***n*** is the generic term representing all the measurements made.

In order to calculate the mean, take the total of ***all*** measurements, ***X,*** divided by the number of measurements, ***n***; the frequency is the total times a specific measurement appears in the data set.

To get the total of all measurements, each frequency in the table is multiplied (shown by the symbol *) by each ***X***.

Then the sum of (X * Frequency) gives the total measurement. The symbol for "sum of" is **Σ** and the pair of symbols "**Σ*X*"** is read as the "***sum of X***."

X	FREQUENCY	X * FREQUENCY
1	5	5
2	3	6
3	4	12
4	1	4
5	3	15
6	1	6
7	4	28
8	2	16
Total	n = 23	Σx = 92

Figure 39: The total number of items in the data set is X times the frequency.

Total: ***Σ(X * Frequency)*** = ***92***

Number of measurements **n = 23**

The symbol for the population mean is μ, the Greek letter "mu" (lower-case Greek m) pronounced "mew."

The symbol for the sample mean is $\bar{x}$.

The formula for the mean written in symbols is:
$\bar{x} = \Sigma X \div n$ or $\Sigma X/n$ where **/** is "divide by" and
"**ΣX/n"** is read as "the sum of ***X*** divided by **n**."

Calculating the mean:
Mean = 92/23 = 4

DETERMINING THE MEDIAN:

You probably are already aware of the term "median" as used to describe the strip of ground between opposite directions on a freeway. Median means middle. When working with a dataset, the median is the middle number in the set of ***ordered*** measurements. It is essential to order the data in either ascending (smallest to largest) or descending (largest to smallest) order. If you have ***an odd number*** of measurements, there is only one median or middle number; if there is ***an even number*** of measurements there are two middle numbers. By convention the median falls halfway between these two numbers, that is, in either case, the median may be defined as the midpoint of the data set.

EXAMPLE 1:

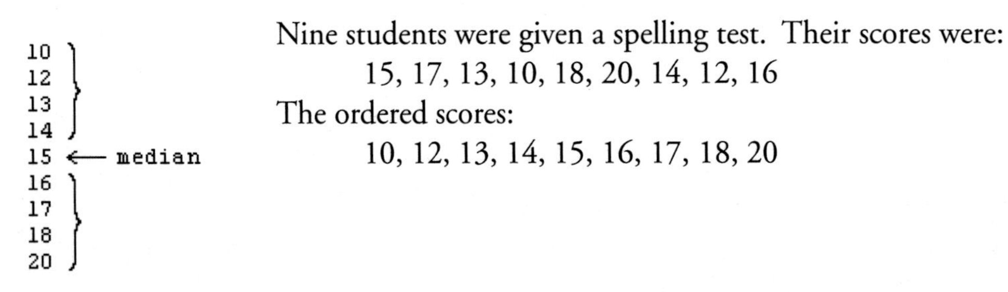

Nine students were given a spelling test. Their scores were:

15, 17, 13, 10, 18, 20, 14, 12, 16

The ordered scores:

10, 12, 13, 14, 15, 16, 17, 18, 20

Figure 40: The location of the median when there is an odd number of scores.

There are nine scores, so the middle score (median) is the 5th one counting from either end (15). Some people feel that listing the score in a column makes it easier to determine the median. This dataset has an odd number of scores, so finding the median is simple.

EXAMPLE 2:

10
12
13
14
← median
17
18
19
20

Suppose the student who got full marks (20) on the previous test, is allowed to skip the next spelling test. Here are the second test scores of the other eight students:

10, 18, 20, 14, 17, 19, 12, 13

Ordering the scores:

11, 12, 13, 14, 17, 18, 19, 20

Figure 41: The median when there is an even number of scores.

Since there is an even number of scores, the middle or midpoint of the data-set is between 14 and 17. With an even number of scores, the median is the average of the two middle scores. If we arrange the scores vertically, we get. There are no specific scores at the middle point, so it must be calculated.

Median = (14 + 17)/2
= 31/2
= 15.5

EXAMPLE 3:

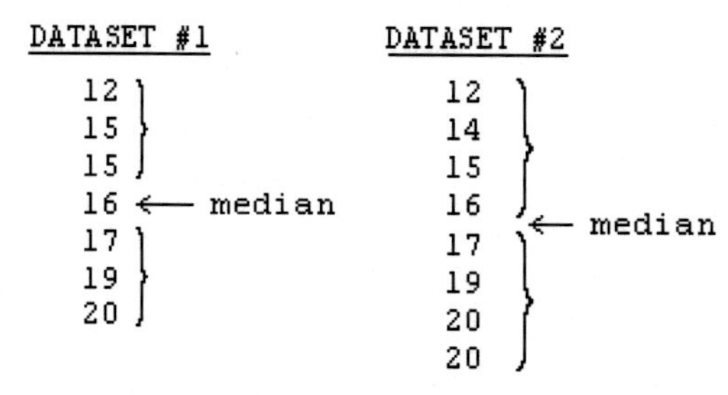

Figure 42: Finding the median in these two data sets.

In the previous two examples, all scores were unique. Suppose we had scores that were duplicated? Again the first step is to place the scores in an ordered array treating each one separately.In this example we will use two datasets, one with an uneven number of scores, the second with an even number of scores.

While the median in Dataset #1 is clear, we can only tell by observation that in Dataset #2, the median is between 16 and 17.

As with the earlier example, sum the two scores and divide by 2.

(16 + 17)/2 = 33/2 = 16.5.

SUMMARY FOR CALCULATING THE MEDIAN:

In determining the median, even if there are duplicated scores, treat each score as if it were unique. Dataset #1 is easy; there are an odd number of scores so the median is the middle score (16). Dataset #2 has an even number of scores so take the average of the two middle scores: (16 + 17)/2 = 33/2 = 16.5.

THE ALGORITHM:

1. Place the scores in an ordered sequence.
2. The median is determined by:
   - The middle score if there are an odd number of datapoints, and
   - The average of the two middle scores if there is an even number of datapoints.
3. The middle score or middle pair of scores has an equal number of scores above and below.

X	FREQUENCY
30	5
31	7
32	8
33	10
34	12
35	15
36	19
37	20
38	18
39	17
40	19
41	21
42	25
43	13
44	10
45	16
n = 16	

Figure 43: When there is a large number of scores.

EXAMPLE 4:

Using the Calculated Mean and the Cumulative Frequency:

When we have data in a frequency table and the scores are repeated, we can use the frequency's ***cumulative total*** to determine the score in which the midpoint occurs. Using the cumulative total allows you to ignore whether there is an even or an odd number of scores. For example, 225 students were asked to respond to an attitude questionnaire. The lowest possible response was 30, the highest was 45. Which score has the median?
When the scores are ordered, you can determine the median ($\Sigma X/n$ for even number of scores; or $\Sigma X/n$ +

0.5 for an odd number of scores). First however, you need to calculate the actual frequency of each score. The following frequency tables were prepared. Note that the third column in the left-hand table becomes the second column in the right-hand table, eliminating the frequency column.

Median:

X	FREQUENCY	X*FREQUENCY
30	5	5
31	7	7
32	8	8
33	10	10
34	12	10
35	15	12
36	19	13
37	20	15
38	18	16
39	17	17
40	19	18
41	21	19
42	25	19
43	13	20
44	10	21
45	16	25
n = 16		ΣX = 225

X	X*FREQUENCY	CUMULATIVE Fr
30	5	5
31	7	12
32	8	20
33	10	30
34	10	40
35	12	52
36	13	65
37	15	80
38	16	96
39	17	114←113
40	18	
41	19	
42	19	
43	20	
44	21	
45	25	
n = 16		

$\Sigma X/n + 0.5$
= 225/16 + 0.5
= 113
Choose the frequency containing 113.
The score with the median is 39.

Figure 44: A cumulative frequency is helpful.

EXAMPLE 5:

Sometimes frequencies are "scattered" and one or more scores ay have zero. There are two ways to determine which score contains the median.

FIRST METHOD:

SCORE	FREQUENCY	CUMULATIVE Fr
12	0	0
16	0	0
18	0	0
24	0	0
10	1	1
11	1	2
20	1	3
25	1	4
13	2	6 ←
15	2	8 ←
19	2	
21	2	
14	3	
17	3	
22	3	
23	4	
n = 12	25	

Figure 45: Working with unsorted scores.

Do ***not*** sort the frequency.

The median is ***n = 12/2 = 6.***

However, since there is an even number of scores (12), we are looking for the ***6.5th*** score for the median.

In the cumulative frequencies, the score ***14*** contains the median. Stop adding when the cumulative total is equal to or greater than the median (6.5).

This may seem to be an unusual result because the score 14 is located elsewhere. The median is still found between the scores of 13 and 15. This is because the median is determined by the score NOT by the frequency; the frequency gives the position of the median.

Not sorting the scores can cause confusion. It is recommended that you always sort the scores

SECOND METHOD:

SCORE	FREQUENCY	CUMULATIVE Fr
10	1	1
11	1	2
12	0	2
13	2	4
14	3	7 ←
15	2	
16	0	
17	3	
18	0	
19	2	
20	1	
21	2	
22	3	
23	4	
24	0	
25	1	
n = 12	TOTAL = 25	

Figure 46: Working with ordered scores (same data as Figure 44)

For this example, order the frequencies; which means that the scores must be ordered correspondingly.

There is an even number of scores out of 12, which means the ***6.5th*** score is the median.

The score, holding the median, lies between 13 & 15 (i.e. greater than 13, but less than 15) or 14.

For simplicity, when determining the median, choose the second method with so that scores are sorted in proper order.

EXAMPLE 6:

There are times when your data needs to be sorted into frequency intervals.

INTERVAL	FREQUENCY	CUMULATIVE Fr
0 - 5	1	1
6 - 10	2	3
11 - 15	5	8
16 - 20	10	18 ←
21 - 25	8	
26 - 30	4	
30 - 35	2	
36 - 40	1	
41 - 45	0	
46 - 50	1	
	n = 34	

Figure 47: Scores are sorted into frequency intervals.

Suppose there are ***34*** Scores. First sort your data so that the intervals are consecutive.

The median is in the fourth interval, that is, between the ***17th*** and ***18th*** scores.

The interval holding the median is ***16-20.***

WHEN TO USE MEDIAN OR MEAN:

In part, this is determined by the data you have. If you need to do further analysis of your data in order to decide whether you can reject the null hypothesis, you will need the mean. Otherwise, examine your data: If you have a single ***outlier*** (datapoint outside the range of all other data) the median will give a better measure of the central tendency. If the sample is large and does ***not*** include any outliers, then the mean is a better measure of central tendency.

THE MODE:

The term "mode" has been used in the field of fashion for someone who is in the highest of currently popular fashion. If a shoemaker is interested in finding the most popular size of shoe, he is looking for the mode. In analyzing data, the ***mode*** is defined as the value or the interval value which occurs most frequently. There can be more than one mode and the mode may not be the central value (median) in a distribution. ***There is no formula for calculating the mode.***

EXAMPLE 1:

DAY OF THE WEEK	# OF ITEMS SOLD
Monday	375
Tuesday	468
Wednesday	572
Thursday	483
Friday	645 ←
Saturday	587

Figure 48: Friday is the busiest day.

The data must be scanned to determine which category contains the highest frequency as the highest frequency gives the mode. ***Note:*** while the frequency determines which category is the mode, ***the mode is the category not the frequency.***

The data to the right shows the number of books checked out each day, during one week of the semester. We can see that the mode is Friday.

Conclusion: The total number of sales shows that Friday is the busiest day of the week (the Mode).

EXAMPLE 2:

SECTION	# OF STUDENTS
1	119
2	123 ←
3	107
4	120

Figure 49: Section 2 has the most students.

The following is data for Professor X, showing him the number of students registered in the four sections he will teach during the coming semester.

Section 2 contains the greatest number of students.

Conclusion: Section 2 has the highest number of students registered, i.e., it is the most popular class.

MORE THAN ONE MODE:

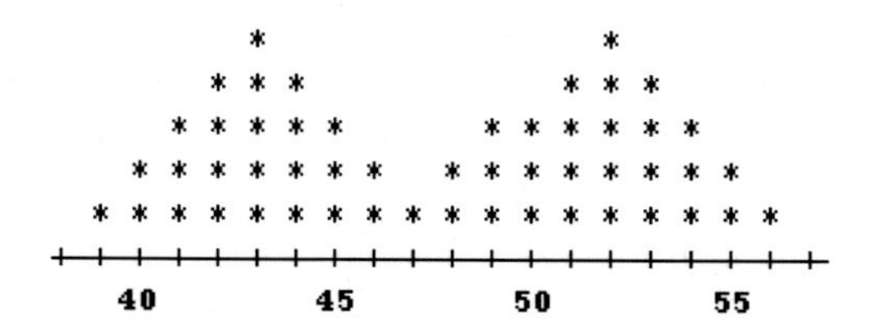

Figure 50: A data set with two modes.

Although it is fairly unusual, data may have more than one mode. If there are two distinct peaks in the graph even though they may not be exactly the same height, they should be close. The data is ***bi-modal*** unless there is a large difference in the height of the two peaks.

It is also possible to have data that is ***tri-modal*** (three peaks all of a similar height). Not all data show a mode. When there are several "peaks" in a graph (more than three), despite the fact that all are the same or close to the same height, there is no mode.

SUMMARY - DIFFERENCES BETWEEN MEAN, MEDIAN, AND MODE:

This unit discussed the three measures of Central Tendency, the ***mean***, the ***median*** and the ***mode***.

THE MEAN: is the most widely used of these measures in analyzing research data and forms the basis for a number of other testing procedures, including the ***standard deviation***, the ***z-score***, and the ***t-value***. To calculate the mean, the total frequency (the sum of the frequency of each score ***X*** times the value of ***X***) is used. The mean can be regarded as the balance point, or center of gravity of a data distribution. On either side of the mean, there is an equal "weight" of data.

THE MEDIAN: is the middle value in the distribution, if there are an odd number of scores. If there is an even number of scores, divide by 2 and add a half point. If there were 16 scores, the median would be (16/2 + 0.5) or 8.5.

THE MODE: is the score with the greatest frequency. Some distributions have two or three modes, where the highest frequencies of scores are very similar or the same.

UNIT 3C-1 Assignment:

1. Describe the difference between the terms: the mean, the median, and the mode and how each one is calculated.

2. Given data in the form of a frequency distribution, calculate the mean and determine the median and mode.

STUDENT	EXAM SCORE	(SORTED) STUDENT	SORTED EXAM SCORE
J.A.	75		
K.C.	90		
A.G.	95		
M.P.	75		
P.T.	60		
F.W.	70		
	n = 6		n = 6

Figure 51: Data for Assignment Question 2.

UNIT 3C-1 Assignment Feedback:

Q2. Given data in the form of a frequency:

STUDENT	EXAM SCORE
J.A.	75
K.C.	90
A.G.	95
M.P.	75
P.T.	60
F.W.	70
	n = 6

(SORTED) STUDENT	SORTED EXAM SCORE
A.G.	95
K.C.	90
J.A.	75
M.P.	75
F.W.	70
P.T.	60
	n = 6

Figure 52: The frequency table for Question 2 has been completed.

n = 6

Mean = ΣX/n

= (95+90+75+75+70+60)/6

= 480/6

= 80

Median: *There are an even number of observations so the median is the 3.5th score. But there are two scores both 75. So the median is 75.*

Mode: *is the highest score of 95 and A.G. has the highest score.*

UNIT 3C-2: MEASURES OF SPREAD - RANGE, VARIANCE & STANDARD DEVIATION[7]

PURPOSE:

The variance and standard deviation form the basis for several of the tests for interval data. Both measures determine the spread of data around the mean. For this reason, the measures are treated separately from any of the statistical tests.

OBJECTIVES:

- Define the terms variance and standard deviation, and describe how they relate to the mean and to a normal curve.
- Given data in the form of a frequency distribution, calculate the mean then use it to calculate the variance and the standard deviation.
- When you have collected all your data, calculate the Variance, and Standard Deviation for **your** data.

7 Useful Internet Reference: http://stattrek.com/AP-Statistics-1/Variability.aspx?Tutorial=stat

INTRODUCTION:

One problem with the mean is that, although it shows how close data is to the central point of the curve, it does not show how far away, that is, how dispersed the data is from that central point. One measure of the spread of data is the ***range.*** This was discussed briefly in ***Unit 3B:*** *Frequency Distributions.* The range is the difference between the highest and lowest scores. But that gives limited information, particularly when you have a large spread of data. All the range does is show where the beginning and end of the dataset occurs. If you are using class intervals, the range is the highest and lowest datapoints in the two outer class intervals.

The mean, however, is useful as it can be used to calculate the ***variance*** and ***standard deviation*** which are measures of how far data is from the mean. The variance and standard deviations give much better information. But before discussing these, there is one important aspect of collecting data that you need to watch for: extreme data or ***gross outliers.*** These can distort the data as a gross outlier lies outside the normal range of data.

THE RANGE & GROSS OUTLIERS:

Suppose a class is available for student to learn how a laptop computer is used in the university library. In addition to the undergraduates, the local high school has enrolled one of their students in the class. The range is the difference between the highest and lowest datapoints.

HIGH SCHOOL STUDENT'S DATA INCLUDED:

Raw data:	15, 28, 30, 35, 32, 27, 34.
Sorted:	15, 27, 28, 30, 32, 34, 35.
Mean:	(15 + 27 + 28 + 30 + 32 + 34 + 35)/7 = 169/7 = 24.14
Range:	35 - 15 = 20 years

HIGH SCHOOL STUDENT'S DATA NOT INCLUDED:

Raw data:	28, 30, 35, 32, 27, 34.
Sorted:	27, 28, 30, 32, 34, 35.
Mean:	(27 + 28 + 30 + 32 + 34 + 35)/6 = 154/6 = 25.67
Range:	35 - 27 = 8 years

The high school student is an "odd man out," and by including his/her age in the calculation, the range is skewed as a result. While in this example the means are fairly close together, the range gives an incorrect impression about the "normal" makeup of the dataset. The inclusion of the high school student produces a number line datapoint plot like this:

If the 15 year old is excluded, the left side of the graph line before 25 is eliminated and the right side of the graph line from 25 is left giving a totally different picture.

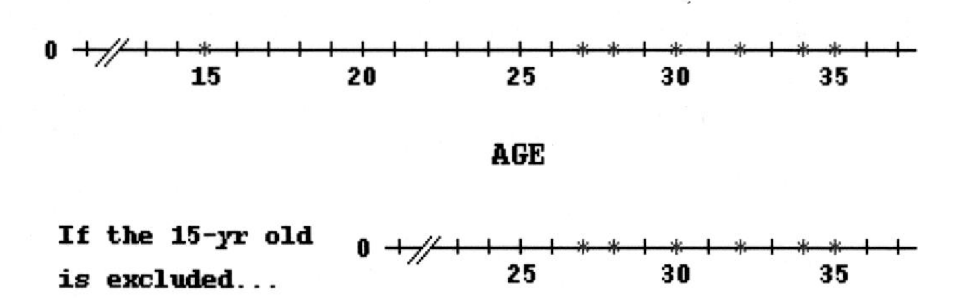

Figure 53: Comparing data with an outlier and when the outlier is excluded.

Extreme data at either end of a dataset is always going to skew the results. But you have a dilemma - do you include that datapoint to get a complete picture or exclude that datapoint to get a more realistic picture? Then again, what if you have gross outliers at both ends of a range of data? Will they balance each other or what? ...

EXPLAIN WHY AN OUTLIER EXISTS:

With any gross outlier, it is wise to find out ***why*** that datapoint exists. Go back to the source of your data and find out if there are possible reasons for the "far out" datapoint. It may have an impact on your report.

For example, in the above computer class: Does the high school send students to each one of the possible library computer classes and it just so happens that only one ended up in this class? Or perhaps the high school student was an experiment to see how such students would "fit in" to a university class? Or perhaps we have a very bright student who, despite the age, is a high school senior and will be attending university full time next year? In your data, it is possible that the outlier datapoints can add insight to your report.

THE VARIANCE:

Will calculating the ***variance*** and the ***standard deviation*** give us a better picture of the data? The definitions of several terms relating to variance are helpful:

VARIABILITY:

SCORE (X)	FREQUENCY	X times FREQUENCY
1	5	5
2	3	6
3	4	12
4	1	4
5	3	15
6	1	6
7	4	28
8	2	16
9	2	16
	n = 23	$\Sigma x = 92$ $\bar{X} = 4$

Figure 54: Use X*frequency to calculate the mean, where the symbol * represents multiply.

Definition: Something liable to vary (differ) in some characteristic. Suppose you had a random sampling of students in six undergraduate classes; the same topic taught and the same professor reduces these two factors' variability to a minimum. The instructor's presentations while similar may vary slightly. Good instructors will adapt the content of the topic as they learn from each presentation. So there may be more variability than expected.

Suppose you sampled three students from each of the six classes and checked the ages of each student. If there is an average of ten students in each class, similar ages should result and variability probably is small. However, if there's an average of 100 students in each class, ages within each sample would probably be more variable. Sampling only three students out of 100 the chance of variability in your data is greatly increased. ***When preparing to do a research study, consider the possible effects of variability.***

MEAN:

To calculate the mean we must take into consideration ***every*** measurement, not just the scores. Remember, for calculating the mean, the actual scores (***X***) are not used but the frequency multiplied by ***X.*** For example, 3, one of the scores in Figure 53, must be multiplied by its frequency of 4 to get the total of 12. (Think: three people had a score of 4, so the total scores of these three are 3*4 where * is the symbol for multiply.) The sum of all observations (**Σ*X***) is 92.

The mean ($\bar{x}$) is **Σ*X*/n**, which is 4.

VARIANCE:

The variance is a way of measuring the variability of the data to determine how different from the mean each observation is. As the mean is near the middle of the observations, and we are measuring the values of ***X*** compared with the mean, some of the measurement are located above (greater than the mean) and some below (less than the mean). As a result, the deviations have positive and negative values. The third column in Figure 54 is titled "Differences" since this is only the first step in calculating the total deviations. Total deviations must be found by multiplying each of the differences by ***X***, the frequencies for each difference.

SCORE (X)	FREQUENCY (Fr)	DIFFERENCES	TOTAL DEVIATIONS
1	5	1 - 4 = -3	5 * -3 = -15
2	3	2 - 4 = -2	3 * -2 = -6
3	4	3 - 4 = -1	4 * -1 = -4
4	1	4 - 4 = 0	1 * 0 = 0
5	3	5 - 4 = +1	3 * +1 = +3
6	1	6 - 4 = +2	1 * +2 = +2
7	4	7 - 4 = +3	4 * +3 = +12
8	2	8 - 4 = +4	2 * +4 = +8
	n = 23 $\bar{X}$ = 4		-15-6-4-0=-25 +3+2+12+8=+25

Figure 55: Differences are used to get the total deviations.

Column 3: The differences are found by subtracting the value of the mean ($\bar{x}$) from each score (***X***). In this example, $\bar{x}$=***4***.

Column 4: Again, since you wish to account for all observations, multiply the differences by the frequencies. This will give you the ***total deviations.*** One way you can tell if you have calculated correctly is that the sum of the deviations will be zero. [-25 plus +25 equals 0]. The positive values cancel out the negative values.

For ***every*** dataset, the sum of the frequencies times the deviations ***always*** equals zero. If they do not, your calculation is incorrect. In Figure 55, the ***results*** of the calculation in column 4 of Figure 54 are the only information that is transferred.

CALCULATING THE VARIANCE:

What we are trying to do is calculate the ***variance.*** But how do we deal with something that sums to zero? To get around that, we ***square*** the deviations. This also has the advantage of removing the negative signs. Calculations are to two decimal places.

SCORE	FREQUENCY (f)	DIFFERENCES	DEVIATIONS $X - \bar{X}$	$(X - \bar{X})^2$
1	5	-3	-15	$(-15)^2 = 225$
2	3	-2	-6	$(-6)^2 = 36$
3	4	-1	-4	$(-4)^2 = 16$
4	1	0	0	$(0)^2 = 0$
5	3	+1	+3	$(+3)^2 = 9$
6	1	+2	+2	$(+2)^2 = 4$
7	4	+3	+12	$(+12)^2 = 144$
8	2	+4	+8	$(+8)^2 = 64$
	n = 23		-25+25 = 0	$\Sigma(X - \bar{X})^2 = 498$

Figure 56: Multiply the differences by the frequency.

In this example, the variance is a single number that is the average of the squares of each deviation from the mean then by dividing the sum of the squares by n-1 [n-1 = 22].

So the variance is 498/22 = 22.64

The algorithm (a step-by-step set of instructions) to calculate the variance is as follows (see Figures 53, 54, and 55):

- Calculate the mean ($\bar{X}$).
- Divide the sum of all measurements (ΣX) by the number of observations ($\Sigma X/n$).
- Calculate the difference from the mean for each measurement in the dataset.
- Calculate the deviation by multiplying each difference by the frequency of each and every **X; ($X-\bar{X}$)**
- Square each deviation $(X-\bar{X})^2$
- Sum the squares of all deviations $\Sigma(X-\bar{X})^2$
- Divide the result of step 6 by one less that the total number of measurements in the set **(n-1)**. When you divide the variance by **(n-1)** rather than **n** the result is closer to the population's variance.

STANDARD DEVIATION:

To calculate the mean we must take into consideration ***every*** measurement, not just the scores. Remember, for calculating the mean, the actual scores (***X***) are not used but the frequency multiplied by ***X.*** For example, 3, one of the scores in Figure 54, must be multiplied by its frequency of 4 to get the total of 12. (Think: three people had a score of 4, so the total scores of those three are 3*4 where * is the symbol for multiply.)

We look for a measure that will indicate the deviation for the whole set of observations. That is, a quantity calculated to indicate the amount of deviation for the set as a whole and is called the ***standard deviation.***

DEVIATION:

The deviation is the distance between any single measurement and the mean of the dataset. The mean is the sum of ALL observations divided by n.

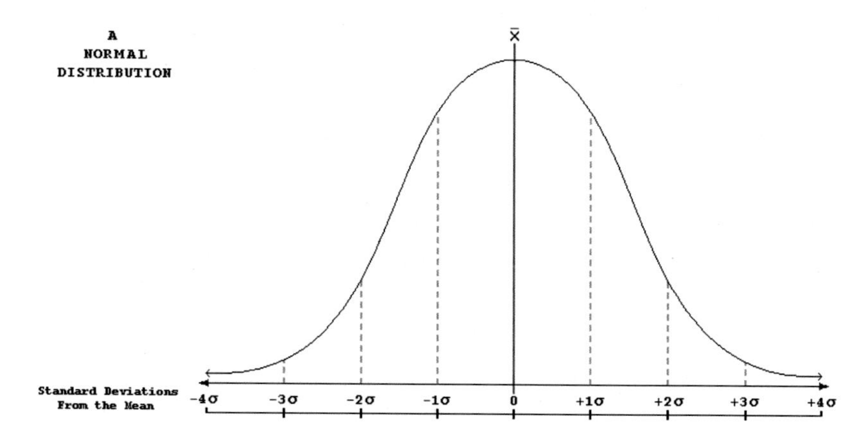

Figure 57: In a normal distribution, the standard deviation σ of the population forms equal-width divisions across the curve as shown.

To find the deviations you subtract the mean (***0***) from the actual scores (***X***). So the ***mean*** is based on finding the sum of ***all*** scores or observations in the dataset. The ***deviations*** are based on the difference between the mean and each unique individual score.

When we work with statistical measures we often try to "standardize" them. Look at Figure 53; all the deviations are measured from the mean and have been standardized.

The letter S_x represents the ***standard deviation of a sample*** where ***X*** represents the observations of the sample. The ***population***'s standard deviation is represented by the lower case Greek letter "sigma" or **σ** (see Figure 56).

The standard deviation is based on the variance. Now, the standard deviation is not a square measure, so to calculate the standard deviation we must take the square root of the variance. For the variance calculated on page 154 (see VARIANCE): 498/22, or 22.64 to two decimal places, the standard deviation is the ***square root*** of the variance.

$S_x = \sqrt{22.64}$
$= 4.76$ (to 2 decimal places)

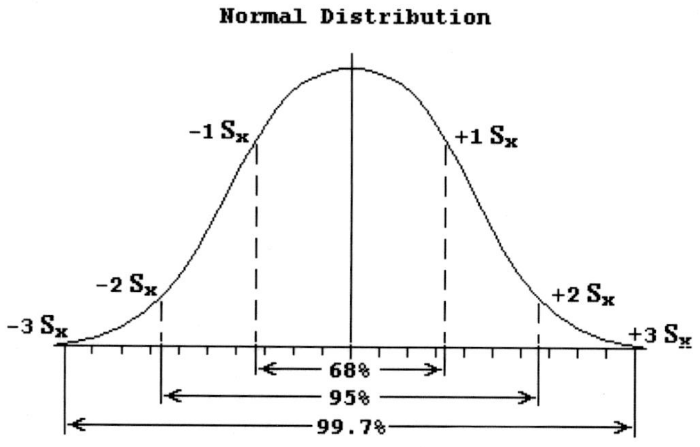

Figure 58: A normal distribution showing the empirical rule for the standard deviation, S_x.

THE EMPIRICAL RULE:

What does it mean when you have a standard deviation? There is a rule called the ***empirical rule*** concerning standard deviations (S_x). Empirical refers to information or decisions based on observations rather than on theory or pure logic.

If the frequency curve of your data is symmetrical and bell-shaped, then about ***68%*** of all measurements fall within one standard deviation of the mean, 95% would be within two standard deviations and 99.7% within three standard deviations (shown by the dashed lines).

Described in a different way: on a normal curve with 100 datapoints: 68 measurements would be within one S_x (standard deviation) of the mean, 95 measurements would be within two S_x, and at least 4 measurements (99.7%) would be would be within three S_x.

Reminder: positive values are to the right of the mean and negative values to the left of the mean.

THE POPULATION STANDARD DEVIATION (σ) :

The population standard deviation (***sigma***) would be preferable to a sample standard deviation. However, sigma is rarely known. As a result, we have to use the sample standard deviation. As a general rule, if S_x is low in value, the datapoints are close to the mean. If S_x is high, the datapoints are spread out; the higher the value of S_x, the greater the spread or dispersion of the datapoints.

ALGORITHM FOR CALCULATING THE STANDARD DEVIATION:

The ***standard deviation*** is a more dependable measure of variability than the range of the data because it uses every observation, rather than only the two extremes. It is usually written as S_x to show that it is the standard deviation of all measurements of ***X.*** There is only one method for calculating the standard deviation: Take the square root of the variance. Therefore, from the algorithm for the variance, it is only necessary to add one more step, the seventh step below to the algorithm for finding the variance:

1. Calculate the mean ($\bar{X}$).
   - Multiply each and every **X** by the frequency. For the calculation of the mean, this becomes **X**.
   - Identify the total number of observations **(n)** in the dataset.
   - Find the sum of all measurements **(ΣX)** in the set.
   - Divide the sum by the number of observations **($\Sigma X/n=\bar{X}$)**.
2. Calculate the difference from the mean for each measurement in the dataset.
3. Calculate the total deviations by multiplying each difference by the frequency of each and every **X; ($X-\bar{X}$)**
4. Square each deviation; **$(X-\bar{X})^2$**
5. Sum the squares of all deviations; **$\Sigma(X-\bar{X})^2$**
6. Divide the result of step 6 by one less that the total number of measurements in the set; **(n-1)**. When you divide the variance by **(n-1)** rather than **n** the result is closer to the population's variance.
7. Take the square root of the variance to get the standard deviation: **$\sqrt{[\{\Sigma(X-\bar{X})^2\}/(n-1)]}$**

EXAMPLE USING THE ALGORITHM:

An example in step-by-step and table form shows the process. Ten books first published in 1950 and regularly used by the psychology department were studied. The topic of interest was the number of editions of each book that had been published between 1950 and the current year (20xx).

STEP 1:

BOOK ID	EDITIONS (X)
1	4
2	3
3	10
4	5
5	2
6	3
7	6
8	7
9	12
10	8
	n = 10 $\bar{X}$ = 6

Figure 59: When the Book ID is used to sort data, the frequency, the number of editions, is left unsorted.

- Identify the number of observations (n) in the set.

  n = 10

- Find the sum of all measurements (**ΣX**) in the set.

  ΣX = 4+3+10+5+2+3+6+7+12+8

  = 60

- Find the mean ($\bar{X}$) of the set of datapoints.

  = 60/10

  = 6

RECOMMENDED: Always sort your scores ***before*** starting the calculations especially if the only other information available is nominal such as the Book ID.

The frequencies have not been sorted in this illustration. Is it truly necessary to do so?

DATA SORTED BY FREQUENCY:

BOOK ID	EDITIONS (X)
9	12
3	10
10	8
8	7
7	6
4	5
1	4
2	3
6	3
5	2
	n = 10 $\bar{X}$ = 6

Figure 60: Data has been sorted by frequency.

Compare Figure 58 above with its unsorted frequencies and Figure 59 here, where the frequencies have been sorted. The Book IDs have been matched to the appropriate sorted frequency (***X***). Notice that the differences now show a pattern.

Rather than sorting the data by the number assigned to each book, (an arbitrary nominal value, that is, a label), sort the scores (***X***). This will enable you to see the pattern of the differences much more effectively.

Some people believe that sorting raw scores means the scores are no longer "raw." But the raw score values are unchanged; the scores have simply been placed in a more useful order. The benefit of sorting the scores: it is easier to see any patterns that develop when the rest of the calculations are completed.

BOOK ID	EDITIONS (X)	DIFFERENCES
1	4	4 - 6 = -2
2	3	3 - 6 = -3
3	10	10 - 6 = 4
4	5	5 - 6 = -1
5	2	2 - 6 = -4
6	3	3 - 6 = -3
7	6	6 - 6 = 0
8	7	7 - 6 = 1
9	12	12 - 6 = 6
10	8	8 - 6 = 2
	n = 10 $\bar{X}$ = 6	

BOOK ID	EDITIONS (X)	DIFFERENCES
9	12	12 - 6 = +6
3	10	10 - 6 = +4
10	8	8 - 6 = +2
8	7	7 - 6 = +1
7	6	6 - 6 = 0
4	5	5 - 6 = -1
1	4	4 - 6 = -2
2	3	3 - 6 = -3
6	3	3 - 6 = -3
5	2	2 - 6 = -4
	n = 10 $\bar{X}$ = 6	

Figure 61: Comparing the effect of sorting on the pattern shown by the calculated differences.

Figure 60 also shows the problem with using numerals as labels. One's mind is conditioned to see numbers for their integer values. When labeling data, it is much better to either use symbols or alpha characters as shown in Figure 61. In this illustration, the alpha characters have been placed without regard to the order of the numerical characters.

It is irrelevant what order the nominal IDs are placed as the characters are only used as a means of identification. It is recommended that the Book IDs be assigned at any time in the collection of data as long as the assignment is made BEFORE making any calculations. For consistency and accuracy, do not rearrange the assigned IDs once you begin analyzing data.

STEP 2:

BOOK ID		EDITIONS (X)	DIFFERENCES
9	A	12	12 - 6 = +6
3	B	10	10 - 6 = +4
10	C	8	8 - 6 = +2
8	D	7	7 - 6 = +1
7	E	6	6 - 6 = 0
4	F	5	5 - 6 = -1
1	G	4	4 - 6 = -2
2	H	3	3 - 6 = -3
6	J	3	3 - 6 = -3
5	K	2	2 - 6 = -4
		n = 10 $\bar{X}$ = 6	

Figure 62: Correcting the labels for the BOOK IDs.

Calculate the differences from the mean for each measurement in the dataset.

STEP 3:

BOOK ID	EDITIONS (X)	$(X-\bar{X})$ DIFFERENCES	$X(X-\bar{X})$ TOTAL DEVIATIONS
A	12	12 - 6 = +6	12 * +6 = +72
B	10	10 - 6 = +4	10 * +4 = +40
C	8	8 - 6 = +2	8 * +2 = +16
D	7	7 - 6 = +1	7 * +1 = + 7
E	6	6 - 6 = 0	6 * 0 = 0
F	5	5 - 6 = -1	5 * -1 = - 5
G	4	4 - 6 = -2	4 * -2 = - 8
H	3	3 - 6 = -3	3 * -3 = - 9
J	3	3 - 6 = -3	3 * -3 = - 9
K	2	2 - 6 = -4	2 * -4 = - 8
	n = 10 $\bar{X}$ = 6		

Figure 63: The total deviations are the frequency times the differences.

Calculate the total deviations by multiplying each difference by the frequency of each and every ***X then subtract the mean.***

You can now see that with the mean being six (6), the difference of 6-6 is zero (Book ID ***E***), and the rest of the differences fall around the mean.

If you were to draw the curve to represent these differences, the positive values would be to the right of the mean and the negative values to the left. While it is not essential to put the plus (+) sign in front of the positive values, it is helpful to do so.

STEP 4:

Square the deviations (column 3) and sum the squared deviations.

$\{\Sigma(X-\bar{X})^2\}$

For emphasis, Column 4 in Figure 62 has become Column 3 in Figure 63.

BOOK ID	EDITIONS (X)	$(X-\bar{X})$ TOTAL DEVIATIONS	$(X-\bar{X})^2$
A	12	12 * +6 = +72	5184
B	10	10 * +4 = +40	1600
C	8	8 * +2 = +16	256
D	7	7 * +1 = +7	49
E	6	6 * 0 = 0	0
F	5	5 * -1 = - 5	25
G	4	4 * -2 = - 8	64
H	3	3 * -3 = - 9	81
J	3	3 * -3 = - 9	81
K	2	2 * -4 = - 8	64
	n = 10 $\bar{X}$ = 6		$\Sigma(X-\bar{X})^2 = 7404$

Figure 64: The variance is based on the sum of the square of each deviation.

STEP 6: CALCULATE THE VARIANCE.

To get the variance, divide the sum by one less than the number of observations (n-1) = 9.

$\{\Sigma(X-\bar{X})^2\}/(n-1)$

= 7404/9

= 822.67 (to 2 decimal places)

= Variance

STEP 7: CALCULATE THE STANDARD DEVIATION:

Take the square root of the variance.

$\sqrt{[\{\Sigma(X-\bar{X})^2]/(n-1)]}$

= √[822.67]]

= 28.68 (to 2 dec pl)

= ***Standard Deviation***

PITFALLS WHEN ADDING GRAPHICS:

Just a few of the possible graphics have been used in this unit. More information on illustrating data is dealt with in Appendix A at the end of Volume 2. Graphics are based on assumptions about the data one collects. Adding graphics to your report can cause real problems if you do not think carefully about your data. You must be totally aware of what type of data you have and its implications as there is a significant differences between interval, ordinal and nominal data.

Book IDs given as numbers tends to suggest that the IDs are interval data. The Book IDs were assigned prior to collecting the data (the number of editions for each book). When the Book IDs are shown as numbers, it is easy (as in not thinking carefully about the data) to graph the data as if it were interval data.

BOOK ID	EDITIONS (X)
9	12
3	10
10	8
8	7
7	6
4	5
1	4
2	3
6	3
5	2
	n = 10 $\bar{X}$ = 6

Figure 65: Numbers were assigned to the books before collecting data.

The result of assuming that the Book ID is interval data is shown below, in Figure 65 where the same data is plotted on a bar graph and on a scatter diagram.

Why do these two graphics cause a problem?

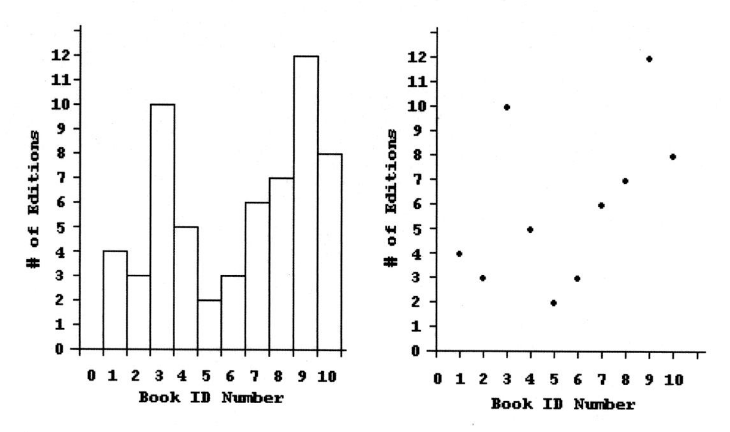

Figure 66: Numerical Book IDs are mistaken for interval data.

The graphing has treated the Book IDs as if they were interval data instead of nominal data. There is no way of drawing a meaningful frequency polygon on the bar graph where there appear to be two modes. Also, the scatter gram shows the datapoints scattered without any observable pattern.

The problem is caused first, by ***arbitrarily*** assigning numbers as the Book IDs ***and second, doing it before*** the data was collected. Because the graphic artist assumed that the data was interval data, a value of zero has been included on the baseline for the Book ID numbers. Not only was no book assigned the number zero (0), but the presence of zero reinforces the idea that the data is interval data when in reality it is nominal.

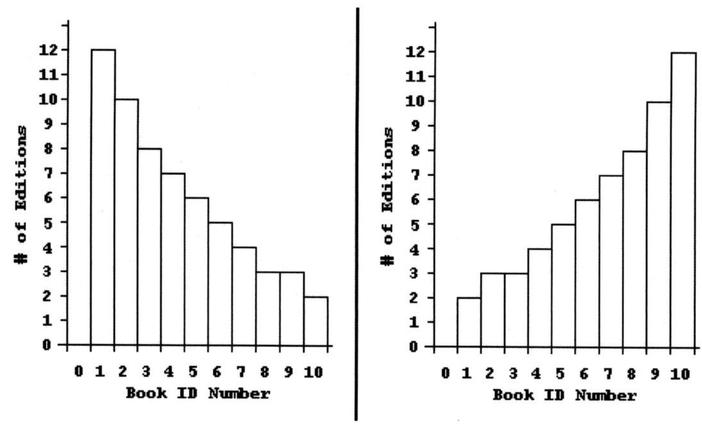

Figure 67: Graphics based on Figure 66, based on the assumption that IDs are interval data.

There would still be a problem if the Book IDs had been assigned after sorting the data but in spite of everything had been assigned as numbers. It would not make a difference if the number of editions (which ***is*** interval data) were ordered from the greatest number or from the least number before assigning the numerical IDs.

Unfortunately, the result of graphing the sorted data in Figure 66 gives what seems to "normal" appearing curve.

RECOMMENDATION:

Never use numerical values if you are working with nominal data, especially if the nominal labels are intended as identification. Use upper or lower case alpha characters.

FINAL EXAMPLES:

These examples, because of larger scores, provide an additional opportunity to calculate the mean and standard deviation. The numerical values in the calculations are much larger than the earlier example. The frequency table is needed to calculate the mean and standard deviation.

FIND THE STANDARD DEVIATION (TO 2 DECIMAL PLACES) & COMPARE THE SOLUTIONS *(Figures 67, 68 & 69):*

First: Scores: 110, 120,123,125, 132,

Since there is only one score for each of the five students, the solution is simple.

1. Calculate the mean (X̄).
2. Subtract the mean from each score to get the deviations (X-X̄)
3. Square the deviations $(X-\bar{X})^2$.
4. Sum all deviations $\Sigma(X-\bar{X})^2$.
5. Divide the sum by **n-1** to get **Variance.**
6. Take the square root of the variance to get Sx the standard deviation [64.5].

STUDENT	SCORE (X)	DEVIATIONS $(X-\bar{X})$	$(X-\bar{X})^2$
A	110	110−122 = −12	144
B	120	120−122 = −2	4
C	123	123−122 = +1	1
D	125	125−122 = +3	9
E	132	132−122 = +10	100
n = 5 n−1 = 4	$\Sigma(X) = 610$ $\bar{X} = 122$	$\Sigma(X-\bar{X}) = -14$ +14 0	$\Sigma(X-\bar{X})^2 = 258$ $S_X = \sqrt{258/4} = 64.5$

Figure 68: The standard deviation is 64.5.

Second: There are ten observations and there are some duplicated. However, they are sorted from smallest to largest, and only one observation is on a line.

STUDENT	SCORE (X)	DEVIATIONS $(X-\bar{X})$	$(X-\bar{X})^2$
A	110	110−123 = −13	169
B	118	118−123 = −5	25
C	120	120−123 = −3	9
D	120	120−123 = −3	9
E	121	121−123 = −2	4
F	123	123−123 = 0	0
G	125	125−123 = +2	4
H	125	125−123 = +2	4
J	132	132−123 = +9	81
K	136	136−123 = +13	169
n = 10 n−1 = 9	$\Sigma(X) = 1230$ $\bar{X} = 123$	$\Sigma(X-\bar{X}) = -26$ +26 0	$\Sigma(X-\bar{X})^2 = 474$ $S_X = \sqrt{474/9} = 52.67$

Figure 69: the standard deviation is 52.67.

The same calculation is possible:

1. Calculate the mean ($\bar{X}$).

2. Subtract the mean from each score to get the deviations **($X-\bar{X}$)**.

3. Square the deviations **($X-\bar{X}$)^2**.

4. Sum all deviations **$\Sigma(X-\bar{X})^2$**.

5. Divide the sum by **n-1** to get **Variance.**

6. Take the square root of the variance to get S_x the standard deviation [52.67].

Third: There are 66 observations, and the frequencies show that there are duplications. In order to account for all observations, (1) the score must be multiplied by the frequencies; and (2) in order to account for all deviations, the deviations must also be multiplied by the frequencies.

FREQUENCIES	(X)	SCORE * FREQ	DEVIATIONS $(X-\bar{X})$	TOTAL DEVIATIONS	$(X-\bar{X})^2$
5	110	110 * 5 = 550	110 − 123 = −13	−13 * 5 = −65	4225
5	118	118 * 5 = 590	118 − 123 = −5	−5 * 5 = −25	625
7	120	120 * 7 = 840	120 − 123 = −3	−3 * 7 = −21	441
8	120	120 * 8 = 960	120 − 123 = −3	−3 * 8 = −24	576
9	121	121 * 9 = 1089	121 − 123 = −2	−2 * 9 = −18	324
8	123	123 * 8 = 984	123 − 123 = 0	0 * 8 = 0	0
7	125	125 * 7 = 875	125 − 123 = +2	+2 * 7 = +14	196
6	125	125 * 6 = 750	125 − 123 = +2	+2 * 6 = +12	144
6	132	132 * 6 = 792	132 − 123 = +9	+9 * 6 = +54	2916
5	136	136 * 5 = 680	136 − 123 = +13	+13 * 5 = +65	4225
n = 66 n − 1 = 65	$\bar{X} \approx 123$	$\Sigma(X) = 8110$ $\bar{X} = 122.88$	$\Sigma(X-\bar{X}) = -26$ +26 0		$\Sigma(X-\bar{X})^2 = 13672$ $S_X = \sqrt{13672/65} = 14.50$

Figure 70: The standard deviation is 14.50.

1. Calculate the mean ($\bar{X}$).

2. Subtract the mean from each score and multiply by each frequency to get the deviations **($X-\bar{X}$)**

3. Square the deviations **($X-\bar{X}$)^2**.

4. Sum all deviations **$\Sigma(X-\bar{X})^2$**.

5. Divide the sum by **n-1** to get Variance.

6. Take the square root of the variance to get S_x the standard deviation [14.50].

UNIT 3C-2 Assignment:

Complete the first three questions and check the results.

1. Describe the difference between the terms: mean, median, and mode and how each one is calculated.
2. Given the frequency distribution below, calculate the mean, median and mode.
3. Use the mean from the frequency distribution to calculate the variance and the standard **deviation.**

When you have finished collecting all your data:

4. Make a report on the data you have collected:
   - Calculate your Mean, Median, Mode.
   - Using your Mean, calculate the Variance and Standard Deviation.
5. When you have the reports from the rest of your discussion group, calculate each of their means, medians, modes, variances and standard deviations to check if their calculations are accurate. If not, circle in red where their calculations are incorrect.
6. Hand in the report, corrected if necessary.

UNIT 3C-2 Assignment Feedback:

1. *Describe the difference between the terms: mean, median, and mode and how each one is calculated.*

   The ***mean*** is the average for the whole set of observations. It is calculated by dividing the sum of all observations **ΣX** by ***n*** the number of observations. **$\Sigma X/n$**

   The ***median*** is the middle observation in a dataset when there is an odd number of observations. If there is an even number of observations, it is the average between the two middle observations. It is best to order the data first.

   The ***mode*** is the observation with the highest frequency. The mode can be ***bi-modal*** or ***tri-modal.***

2. *Given the frequency distribution with the problem, calculate the* ***mean, median*** *and* ***mode. n= 6***

   Mean $= \Sigma X/n$

   $= (100+90+85+75+70+60)/6$

   $= 480/6$

   $= 80$

   Median is between 85 and 75 as there is an even number of observations.

   $= (85 + 75)/2$

   $= 160/2$

   $= 80$

   Mode is K.C.'s score of 100.

3. *Use the mean from the frequency distribution above to calculate the* ***variance*** *and the* ***standard deviation.***

   n-1 = 5

   Variance:

   $\{\Sigma(X-\bar{X})^2\}/(n-1)$

   $= 1050/5$

   $= 210$

   $\sqrt{\{\Sigma(X-\bar{X})^2\}/(n-1)}$

   $= \sqrt{210}$

   $= 14.49$

   = 14.5 to one decimal place

PREVIEW: VOLUME 2

Once you have calculated the standard deviation for any particular set of data, there are five tests that can be applied depending on the type of data you have: The z-test, the t-test, Pearson's correlation test, Spearmen's correlations test and the Chi-Square test. The methods for using these are demonstrated in Volume 2. Algorithms are included.

Printed in the United States
By Bookmasters